Law and Morals

Law and Morals

WARNOCK, GILLICK AND BEYOND

Simon Lee

Oxford   New York

OXFORD UNIVERSITY PRESS

1986

Oxford University Press, Walton Street, Oxford OX2 6DP

Oxford New York Toronto
Delhi Bombay Calcutta Madras Karachi
Petaling Jaya Singapore Hong Kong Tokyo
Nairobi Dar es Salaam Cape Town
Melbourne Auckland

and associated companies in
Beirut Berlin Ibadan Nicosia

Oxford is a trade mark of Oxford University Press

First published 1986 as an Oxford University Press paperback
and simultaneously in a hardback edition

British Library Cataloguing in Publication Data
Lee, Simon
Law and morals: Warnock, Gillick and beyond.
1. Law and ethics
I. Title
340'.112   BJ55
ISBN 0-19-217753-2
ISBN 0-19-283052-X Pbk

Library of Congress Cataloging in Publication Data
Lee, Simon (Simon F.)
Law and morals.
Bibliography: p.   Includes index.
1. Law and ethics. I. Title.
BJ55.L43 1986   340'.112   86-12694
ISBN 0-19-217753-2
ISBN 0-19-283052-X (pbk.)

Set by Katerprint Typesetting Services
Printed in Great Britain by
Richard Clay (The Chaucer Press) Ltd.
Bungay, Suffolk

With love and thanks to Patricia, Jamie, Katie and Rebecca

Contents

Preface

How should the law respond to abortion, discrimination, embryo experiments, pornography, surrogate motherhood, teenage contraception and television violence? I hope that this brief book will encourage constructive, independent thought about such questions of law and morals. It is addressed to the so-called 'general reader'. This mythical character may be a fifth- or sixth-former who considers such issues in General Studies or who is perhaps interested in studying law at university. Another manifestation of the 'general reader' is someone who reads the press coverage of these topics and who would like to think through the underlying issues.

In particular, I have had in mind those university applicants whom I have interviewed in Oxford and London over the past five years. When asked about, say, surrogate motherhood they tend to respond in one of three ways: 'I'm a Catholic so I think it should be banned', or 'I think we should let people do their own thing', or (these are the ones who have been coached) 'it all depends on whether it involves harm to others'. In the space of half an hour, I have enjoyed helping them move beyond such unpromising starts. This book mirrors that process.

At the same time, I hope that the book will meet the wishes of those who have read one or two articles by me in the press and engaged me in further discussion and correspondence.

I have benefited from arguing about these matters with my family, friends, teachers, students and colleagues, especially Ian Kennedy, Joseph Raz and Simon Whittaker. My brother Michael introduced me to the issues when I was a university applicant myself a decade ago. Now he suggested that I write this book. His secretary, Jane Carter, prepared the typescript. Angela Blackburn and Will Sulkin of Oxford University Press saw something worthwhile in the draft. Will's enthusiasm has brought the text rapidly to publication. The editors and publishers of *The Times*, *New Society*, the *Catholic Herald* and the *Law Quarterly Review* gave me permission to incorporate some material which I had first used in their pages.

I am most grateful to all the above for their help. But only I can be blamed for what follows. The speed of the whole process has precluded me from asking colleagues to comment on the text. In any event, the book stands or falls on how it helps 'general readers' to think about law and morals. I would welcome their comments.

Above all, I would like to thank my wife, Patricia, for all her help and encouragement. Our children, Jamie, Katie and Rebecca were all under the age of two at the time of producing the draft and so can be excused for not reading it. The book is dedicated to my family.

The Law and Morals Debate

Sex, drugs and rock-and-roll hit the headlines frequently in the 1960s. Or at least so I am told. They did not dominate my life at St Michael's Roman Catholic Primary School in deepest Kent and I suspect that there was more to life elsewhere in the swinging sixties. But if the media highlighted only some features of that era, the debate among legal philosophers over the enforcement of morality had an even narrower perspective. Never mind drugs or rock-and-roll, discussion seemed to concentrate only on sex.

The famous Hart–Devlin debate[1] centred on whether homosexual practice between consenting adults in private should be decriminalised. Today sex is still an issue for discussion of the relationship between law and morality. In particular, the Gillick case, on the legality of contraceptive advice and treatment for girls under sixteen years old, has revived that theme in the mid-1980s. Yet it was always a narrow perspective. The relationship between law and morality deserves a broader context. There are many more moral issues than those relating to sex. There are many more aspects of the law than criminal prohibition.

Immoral Drugs

In the last two decades, for example, drugs have been much discussed by legislators and judges. Heroin is, at the moment, a problem which judges and the Government are hoping to combat through tougher penalties for drug-pushers. Some people think that softer drugs such as cannabis should be legalised. Others feel that this would encourage a progression to harder drugs and the dreadful effects of addiction. More widely accepted drugs such as alcohol and tobacco have for a long time been subject to legal regulation of one sort or another. One of the most famous and drastic examples in this century is the American experience of prohibition on alcohol, introduced and then ended by constitutional amendments. England in the mid-1980s still regulates licensing hours (though not perhaps for much longer) and the age at which people can lawfully buy alcohol, while drunken driving is a criminal offence.

The state adopts an increasingly equivocal attitude towards cigarette smoking. The law discourages it in many ways: by taxation, age limits and restrictions on advertising. Regulations by public and private bodies restrict smoking in various places. Yet the tobacco industry employs many people, generates much profit and any government would be reluctant, for tax-revenue reasons alone, to ban it altogether. So where do we draw the line? Well, smoking is disapproved of especially where it harms others, hence some bans on smoking in public places. In a sense smoking could be said to harm others indirectly even when it is conducted in the privacy of someone's home since tax-payers contribute to the National Health Service, which might have to care for those whose smoking leads to illness such as lung cancer. On the other hand, the tax on tobacco might cover the cost of dealing with any smoking-related illnesses. Whatever the financial detail, what about liberal principles? However unwise smokers may be, the rest of us may feel we ought to allow them to indulge their habit so long as we are not directly harmed.

So drugs, using the term in the broadest sense to include alcohol and tobacco, raise interesting questions for the law. We may be less likely to think of these matters as questions of morality but if so, it is time to broaden our moral horizons. Whether or not it is right to drink and drive or to take heroin are matters of value judgement, of good and evil, right and wrong, of what we ought to do or ought not to do. This is the stuff of morality just as much as whether homosexuality is as legitimate as heterosexuality.

Immoral Music

Not only drugs but also rock music now raises issues of the enforcement of morality. As I write, the American Moral Majority is pressing for bans on various pop singers whose lyrics and posturing it finds objectionable. What limits should the law place on pop singers' freedom of expression? The question concerns not only the sexual connotations of songs, but also the use of four-letter words and incitement to violence or racial hatred.

Beyond the Criminal Law

Even if we were to confine our attention to sexual morality, we need to broaden our discussion of the enforcement of morality beyond the *criminal* law. The law still treats homosexuals differently from heterosexuals in many important ways. For example, it has been held in a Scottish case that dismissing a man employed at a children's holiday camp because it was discovered that he was homosexual was not an unfair dismissal. The industrial tribunal thought that a 'considerable proportion of employers would take the view that the employment of a homosexual should be restricted particularly when he is required to work in proximity to and contact with children'.[2]

Now, to lose one's job is a far more serious sanction than to be fined £25 for importuning. So while homosexuals are no longer hunted by the criminal law they do not yet receive full protection from our employment law.

Lots of other questions arise in other areas of the law. Should homosexuals be able to adopt children? Should they be able to go through a form of marriage ceremony? If not, why not? Those who would deny homosexuals equal rights argue that society should not, and does not, regard homosexuality as an equally valid alternative to heterosexuality. Some people think that while it is wrong to penalise homosexuals there is no need to grant them equal status. Others would disagree. But in Britain, at least, it is clear that the argument has now moved on from the criminal law question which dominated the debate in the 1950s and 1960s. (In New Zealand, on the other hand, the issue is still whether or not homosexual practice should be decriminalised: see the article in the *Guardian*, 22 October 1985, under the glorious headline, 'Some say yes to gay rights but the others say Sodom'.)

This book will attempt, therefore, to regenerate interest in the relationship between law and the enforcement of morals, developing the issues beyond those raised in the Hart–Devlin controversy. Sex and the criminal law are both very important and both will be discussed here. But there is more to morality than sex and there is more to law than crime.

Legislation

Is there anything more to the whole law and morality debate, however, than a pointless academic exercise? I think that there is. The law affects us. We should aim to affect the law.

Legislators are constantly making decisions about law reform which depend on moral values. We are all capable, albeit to a very slight degree, of influencing Members of Parliament. So we should all be concerned as to how MPs make up their minds on issues of law and morality.

The role of an MP is crucial in voting for a bill, say, to ban embryo experiments or to decriminalise homosexuality or to increase the maximum punishment for drug-pushers. Appropriately enough, issues such as homosexuality, abortion, capital punishment, pornography, video nasties, embryo experiments and surrogate motherhood are not decided along party lines in Parliament. They are instead determined by 'votes of conscience'.

This should not confuse us into believing that only these issues are ones of conscience, as if matters of conscience are not involved in changes in the tax system or the welfare state, or the reduction of inflation, or increasing unemployment, or all the other controversies which take up much more parliamentary time. Most matters before Parliament are matters of conscience, but cynics would say they are only called that when the party whips cannot control their members or where the party leadership does not wish to be seen to be taking a controversial stand. It suits governments that these heated issues are often taken out of their hands by back-bench MPs. More charitably, one could say that the conscience of the MP on other matters determines the political party manifesto which he supports.

How should an MP vote in what is dubbed a question of conscience? One obvious answer is that he should follow his own moral inclinations, hence the term 'vote of conscience'. Another view is that perhaps he might consider what his constituents think. There is at least a plausible argument to be made that MPs should respect the wishes of those who elect them.

Perhaps more influential and more worrying than individual electors' views are the pressures of interest-group lobbying on MPs. We know that different pressure groups have different ideas of morality. The Appendix to the Warnock Report on Human Fertilisation and Embryology lists the organisations which submitted evidence to the Inquiry. Its alphabetical list illustrates the diversity of moral views presented to such committees, beginning as it does with 'Action for Lesbian Parents' and ending with the 'Yorkshire Pro-Life Co-ordinating Committee'. And, of course, it is often committees like Warnock's which do the ground-work before Parliament considers a moral dilemma. If pressure groups want to influence MPs they can start by influencing the committee. Not only Warnock on embryo experiments and *in-vitro* fertilisation but also Wolfenden on homosexual practice and prostitution, and Williams on film censorship and obscenity, illustrate the role of such committees. Cynics would say that their role is to allow the government time in which to do nothing about the problem at hand and observe that their track record in leading to legislation is far from impressive. Nevertheless, perhaps their most important function is to focus public attention on a moral controversy and thereby initiate a 'Great Debate' which itself stimulates and/or educates the public and is beneficial, whatever the legislative outcome.

Personal Reflection

It seems unlikely that citizens have no impact at all on the consciences, or at least the votes, of MPs. But it is only realistic to accept that their significance will vary from the committed to the apathetic and from the influential to the powerless. Nevertheless, the humblest of citizens may still benefit from the 'Great Debate' even if their contributions are ignored by legislators.

When there is the possibility of a change in the law, the media become interested in a moral controversy. When the media become interested, we tend to follow suit. If the newspapers are full of stories about surrogate motherhood (in the wake of the Warnock Report) or teenage contraception (in the light of the Gillick case), our conversations and thoughts are likely to turn to such moral and legal dilemmas.

Anybody who has taught a General Studies or Religious Education class at school will agree that such legal controversies can be the jumping-off point for a discussion about the morality of the conduct in question. Even if the teacher wants to restrict the issue to the question whether it is right for the *law* to enforce a particular moral view, the opportunity is taken to consider the merits and demerits of various moral views on various topics, more or less related to the teacher's concern. This is often illuminating. The 'side-issues' can generate a heated discussion which helps everyone to realise the prejudices and moral views which are held by others and by themselves. (Note the conjugation: I am morally compelled, you are morally misguided, he or she is obstinately prejudiced, and so on.)

Jurisprudence classes can follow the same path. A seminar on the Warnock Report may never get round to embryo experiments but might instead generate a heated debate on whether it is racist for a white couple to want a guarantee that their sperm donor is also white. But such an opportunity for discussion is surely important. So, even if we cannot influence the law, the interest in the subject may help us to clarify our own moral stance. Indeed, that is ultimately more important than the law's position. Even if the law allows us to enter surrogacy arrangements, it is not going to compel us to do so. We have to think through whether we *want* to and one element of that decision will be whether we think it is morally right to do so. Even if the law allows some under sixteen-year-old girls to receive contraceptive treatment, it does not mean that they are all going to demand it. But the Gillick case may have encouraged some young people to think about the wisdom of youthful sexual activity.

A Structure for Thinking About Law and Morals

This book seeks to help readers develop their own views of the relationship between law and morals. In the process, they may also clarify or change their ideas on the moral controversies and, perhaps, on the very nature of morality itself.

We will begin by ridding ourselves of three possible misconceptions. It is sometimes assumed that many people seek to ban everything which they regard as immoral. This mythical body is often called the Moral Majority or the Moral Right. Roman Catholics in Britain are sometimes identified as holding this position. We will explore Roman Catholic attitudes to law and morals in an attempt to show that this is not their position.

The so-called 'Moral Majority' or 'Moral Right' probably thinks of its opponents as the 'Immoral Minority' or the 'Immoral Left'. Whatever the nomenclature, it is sometimes assumed that certain people would never ban anything, however immoral, and would prefer to let everyone 'do their own thing'. We will examine liberal attitudes to law and morals in an attempt to show that this is not the position adopted by liberals (the presumed opponents of the Moral Majority).

It is sometimes felt, particularly by disaffected law students, that much of the law has nothing to do with morality. We will see how many areas of the law in fact reveal a heart of moral values beneath their dry exterior.

Once these misconceptions have been addressed, we will turn to more plausible answers to the question of the proper relationship between law and morals. The traditional response is some variation on a theme by John Stuart Mill, the nineteenth-century philosopher. He maintained that society could only intervene if a person was harming others. We shall see that there are some problems with this approach, such as defining 'harm' and 'others'. We will also consider how two of the most celebrated commentators, Professor Hart and Lord Devlin, have dealt with Mill's principle.

I will then offer what I regard as a more useful analysis of disagreements over law and morals. My approach does not presume to dictate answers, but then I regard that as an advantage. It is, rather, an attempt to separate the elements of dispute, so that we can see exactly what is at issue and how we might structure a decision-making process to resolve the dilemma. It seems to me that there are two reasons for disagreement on controversies over law and morals: first, different predictions as to the factual consequences of legal action, and second, different moral values (or different weights attached to the same moral values).

For example, if X wants to ban surrogate motherhood but Y wants to permit it, their dispute may hinge on:

(1) X's prediction that there will be psychological problems for the child the surrogate mother and the commissioning couple, or that there will be practical difficulties because surrogates will want to keep the baby, as contrasted with Y's prediction that everybody will be delighted with the arrangement or that everybody will keep to the arrangement;

and/or

(2) X's view that it is immoral to involve a third-party in procreation or that it is immoral to use one's womb in this way, as contrasted with Y's view that it is immoral to deny a couple a child which is genetically linked to at least one of them or that it is perfectly moral to use one's womb to nurture a new life.

Of course, this is merely a framework. Plenty of work remains to be done. But if this does help us to clarify the reasons for disagreement, if may be a constructive aid to analysis.

The second part of the book offers a series of contexts within which to explore the relationship between law and morals. We will then see what the harm-to-others test and my preferred approach reveal about various controversies. The controversies themselves have been chosen partly to illustrate the response of diverse decision-making procedures to important moral dilemmas. Committees, courts, administrators or legislators might be faced with the responsibility of attempting to resolve a problem. What are the methods, strengths and weaknesses of each type of decision-maker? In particular, we will devote considerable attention to the Warnock Report, as an example of the way in which an official committee of inquiry operates, and to the Gillick litigation, as an example of judicial reasoning. Readers will also be introduced to the language of such official reports and legal judgments. Rather than break up the flow of the book by giving chapter and verse for each quotation, I have given the references for reports and cases at the end of the text.

But the moral dilemmas have been chosen primarily for their

intrinsic significance. Warnock's concern, the beginning of life and the nature of the family, together with Gillick's issue of contraception, are both related to questions of sexual morality. But the other topics are designed to edge the reader towards wider concerns. We will move from Warnock and Gillick, through other medical issues, to the concepts of equality and liberty. By calling all these issues questions of 'morality', I am clearly adopting a broad understanding of the term 'morality'. As J. L. Mackie has described it, 'A morality in the broad sense would be a general, all-inclusive theory of conduct: the morality to which someone subscribed would be whatever body of principles he allowed ultimately to guide or determine his choices of action'.

Finally, we will conclude with a few morals which can be drawn from the wide-ranging law and morals debate. The readers are left to develop their own ideas with suggestions for further reading, which must constitute a vital part of any introductory book.

Moral Majority?
A Catholic View

The standard approach to the issue of whether the law should enforce morality is to say something like this:

Law is laid down for a great number of people of which the great majority have no high standard of morality, therefore it does not forbid all the vices from which upright men can keep away but only those grave ones which the average man can avoid and chiefly those which do harm to others and have to be stopped if human society is to be maintained, such as murder, theft and so forth.

Who said that? Many people will imagine it comes from John Stuart Mill's famous essay 'On Liberty'. It certainly seems to represent the core of liberal thought. It might be regarded as the sort of permissive approach which annoys moral majoritarians or religious fanatics. But in fact the author of that quotation was St Thomas Aquinas[1] who came up with the idea some 600 years before it occurred to Mill.

There is plenty of support for Aquinas' approach in recent statements by Catholic leaders. Let us first take the case of abortion, in opposition to which Catholics have been at the vanguard of the 'Moral Majority'. Even in their statements on abortion, the church authorities have been keen to emphasise that they do not expect the law automatically to punish all immorality.

Abortion is a special case of 'harm to others' for those who accept that the unborn child is an 'other'. The 'harm' is clearly enormous (destruction in contravention of the 'right to life') and the 'other' is precisely the helpless being who needs the law's protection. As the Vatican Declaration on Abortion maintains, 'It is at all times the task of the State to preserve each person's rights and to protect the weakest'. One might disagree that this is always the task of the State or that the foetus is to count as a 'person'. Nevertheless, it must be acknowledged that the Vatican does not simply leap from the view that something is immoral to the view that it ought to be illegal, without attempting to justify the move.

Indeed that same Vatican Declaration states,

It is true that civil law cannot expect to cover the whole field of morality or to punish all faults. No one expects it to do so. It must often tolerate what is in fact a lesser evil, in order to avoid a greater one.

In order to introduce the British Catholic hierarchy's views on law and morals, it is best to quote from the 1980 statement on 'Abortion and the Right to Life' issued by the Roman Catholic Archbishops of Great Britain. As with the Vatican Declaration, I am not here extracting the argument about whether the unborn child ought to be regarded as having the rights of a born child. I am, rather, presenting the Archbishops' views as an example of reasoning on law and morals. Whether or not the reader accepts the moral position against abortion, the point is to see that the 'Moral Majority' is not unthinkingly expecting others always to be legally subjugated to its moral stance. The Archbishops explicitly state the opposite: 'We do *not* seek to have all Catholic moral teaching imposed by law' (my italics).

Indeed, the following five extracts illustrate various aspects of a well-rounded approach to law and morals: the recognition of conscientiously held contrary moral views; the significance of rights; the wide range of concerns, beyond abortion, which such rights involve; the emphasis on protecting the poor, weak and outnumbered; the provision of practical help for those who reject the option of abortion; the recognition that there remains a moral decision whatever the law; and the point that today's unpopular cause may be tomorrow's accepted truth:

(1) We live in a society where many differing moral and political opinions are conscientiously held and pursued in practice. We make no attempt to override the consciences of our fellow-citizens. We do not seek to have all Catholic moral teaching imposed by law, or even adopted as public policy. But we too have the right, as members of this pluralistic society, to appeal to the consciences not only of our fellow-Catholics, but also of our fellow-citizens and our political leaders and representatives. We too have consciences. And we cannot in conscience remain silent while the most basic human beings are ignored and overridden by the law and, increasingly, by the public policies and everyday practices of our country. These developing human lives may be unborn and silent but they are already our neighbours, living in our midst and are part of our human family. They need to be defended.

(2) Our stand against abortion is one aspect of our stand against all practices that degrade human rights and dignity. Scottish bishops have made many statements, both individually and collectively, on the need to aid developing nations, on social justice at home and abroad, on unemployment problems and on help for the needy and deprived. The bishops of England and Wales issued in 1971 a major statement on moral questions which ranged

over Christian living, race relations, violence and peace. Since then the bishops have tackled the housing problem, disarmament and many current social issues. The bishops have tried to defend the insulted, the despised, the disadvantaged. With other Christians we have resisted racism. We have stressed the brotherhood of man and rejected any discrimination based on colour or race.

The whole of Christian social teaching can be seen as an appeal to the conscience of the relatively well-off and powerful to give practical recognition to the humanity and rights of the poor and the weak. And the social teaching proclaims as well the rights of minorities against majorities who treat them with unfair indifference or hostility.

(3) But when we look at the law of our land . . . we feel obliged to say: unborn children in Great Britain are today a legally disadvantaged class; they are weak; they are a minority . . . Law ought to uphold and embody the principles that are basic to our civilisation and our existing law in every other field; innocent life is to be protected by the criminal law and public policy; no law should countenance discrimination by the strong against the weak.

(4) A girl or woman should always be given the practical help she may need to carry through her pregnancy. She should be given it unstintingly and without moral censure . . . it must always be given in a way that fully respects her freedom and responsibility.

Very necessary and very encouraging are the efforts of those voluntary associations in which Catholics and non-Catholics work together to attack abortion at its root by providing moral and material support to any and every mother-to-be who is willing to allow her baby to be born and not aborted.

(5) To all who are working *against* abortion and *for* the life and future of the unborn and their distressed mothers, we say: do not be discouraged. The laws, practices and opinions of our society may seem, at times, all too firmly set in favour of abortion. Substantial reform may at times seem beyond reach — let alone the full justice which you seek. But your work is not in vain. At the very least it also preserves for everyone an option that would otherwise become stifled and forgotten; the option — of pregnant women and their relatives and friends, of doctors, of nurses, of social workers — to respect innocent life and to refuse to take part in its destruction. And at best, your efforts may well be crowned with success. Success has so often appeared to social reformers to be beyond their reach, almost up to the moment when they attained it.

The Roman Catholic church still seems to assume that outsiders do not believe its claim that it does not equate law and morality. Church leaders therefore are at pains to reiterate their appreciation of the distinction. The Catholic Bishops' Joint Committee on Bio-Ethical Issues, for example, began its evidence to the Warnock Committee with six successive sentences emphasizing the point that they 'do not seek to have all Catholic moral teaching imposed by law'. I wish to explore the reasons behind this approach in an attempt to show liberals that their supposed opponents often agree with them. The immediate response to the Archbishops' Statement might be: why *not* seek to have all Catholic moral teaching imposed by law? A second question would be: *which* parts of Catholic moral teaching should and which should not be imposed by law?

One good reason for restraint is that a law is simply not a suitable instrument for the enforcement of morality in all circumstances. However much the church condemns lustful thoughts the church does not expect the law to intervene, at least until immoral thoughts lead on to immoral actions. Neither is the law the automatic response to all immoral acts. There must come a point at which we realise that the logistics of enforcement, the inordinate use of police and court time and the futility of the possible range of punishments should make us hesitate before leaping from the decision that something is immoral to the conclusion that it should be made criminal.

The second rationale is that the law should be reluctant to override the consciences of those who mistakenly but sincerely believe that they are acting morally. In particular, one should be more reluctant to dictate to others when one's own belief is a matter of faith rather than reason. In those circumstances, by definition, those without 'faith' will not share the same approach to a moral question. They cannot be 'reasoned' into faith. If they are not 'harming others', their consciences should be respected.

A third set of reasons revolves around living in a democracy. Is it right in a democracy to seek law reform when your views are not shared by the majority? In other words, should the Archbishops' Statement be read only in its British context? Should the bishops rightly be more demanding over the enforcement of Catholic morality if the United Kingdom were predominantly Catholic? If so, one could say that the church in the Republic of Ireland might be justified in using its dominant position to entrench the law's refusal to countenance divorce, for instance, but it would be wrong for the British church to attempt the same thing.

Two further, pragmatic, explanations for the bishops' restraint in seeking to reform the law to coincide with morality are sometimes proffered. Firstly, reading the Archbishops' Statement carefully, it does not actually say that they do not *want* their teaching to be imposed by law. They may very well desire just that, but if they feel there is no chance of Parliament agreeing, it makes sense not to *seek* the impossible. By conceding what you cannot expect to achieve you can present yourself as a model of compromise and reason, which may help you at other times. The second ultra-cynical view is that the bishops may be only half-hearted in their commitment to some aspects of the church's teaching. If they were really convinced of the Pope's argument against, say, artificial birth control, would that not affect their desire to incorporate it into the law? Some would argue that our wishy-washy approach may be contrasted with the way in which some Islamic countries merge law and morality in a whole-hearted commitment to their beliefs. Those who accept the principled explanations which I have outlined for distinguishing law and morality can reject such cynicism.

So when the Archbishops say they 'do not seek to have all Catholic moral teaching imposed by law', what does that mean in practice? The example of artificial contraception is a convenient one with which to illustrate the five explanations outlined above. The Roman Catholic church does not seek a criminal law ban on the use of contraceptives by adults even though the famous (but not always understood or even read) papal encyclical *Humanae Vitae* 'teaches that each and every marriage act must remain open to the transmission of life'. One reason is the difficulty and high cost of enforcement allied to the invasion of privacy which such a ban would entail. A second reason is that others disagree with that teaching in all good conscience and the bishops respect their autonomy. A third possibility is that it might seem undemocratic to impose the church's views when many disagree. Fourthly, it would be pointless to try as there is no realistic prospect of Parliament agreeing. Finally, there may be some doubts about whether artificial contraception really is immoral.

But where in the Archbishops' stance on abortion does this principle of restraint feature? The one example I can find in the Archbishops' Statement on abortion where they do not seek to have Catholic moral teaching imposed by law is where the life of the mother cannot be saved without a direct abortion. Here they say that Catholic teaching makes 'exceptionally high demands', perhaps involving 'heroic sacrifice', and so the church does not feel justified in asking the law to punish those who fail to meet such exacting standards. Such cases are, however, 'certainly exceptionally rare or perhaps even non-existent'.

Abortion then may not be the best issue on which to pin the principle of Catholic restraint in enforcing morality through law, precisely because Catholics see it as violating the fundamental rights of the weak. There is no equivocation in the moral stance of the

Archbishops. In the statements already quoted, they begin by saying, 'We speak in defence of life against the evil of abortion'. Why? 'The Church speaks out against abortion, as it has from the beginning, because it acknowledges the human rights and dignity of all, including the unborn, and is committed to their defence. There is here a crucial point of principle. It has everything to do with the intrinsic value and inalienable rights of each individual. It is a matter of respect for our neighbour'. But if one imagines a Catholic dictionary of sin beginning with abortion, adultery and avarice it is at least clear that the Catholic church would not expect all three to be crimes. Adultery illustrates both Catholic restraint in not using the criminal law to support a moral view and a more general point that we need to look beyond the criminal law to appreciate fully society's attitude to immoral conduct. Adultery is not a crime in the United Kingdom but this does not mean that the law regards it as perfectly acceptable behaviour. Adultery constitutes a ground for divorce, a very important part of our civil law with many ramifications affecting the individuals concerned, their children and their property.

As for avarice, we have seen from the Archbishops' Statement on abortion that the church is well aware of avarice causing injustice and the need to counter such inequalities. But, clearly, this is a complex problem which cannot be solved simply by declaring greed to be a criminal offence.

I do not dispute that some Catholics and others *are* inclined to assume that immorality should automatically be met with legal sanctions. Mrs Gillick, for example, seems to hold that view at least in relation to her own crusade.

But the official statements of the Roman Catholic church, in contrast, reflect a more reasoned approach to law and morals. For the moment, then, let us accept that even those who take a strongly religious attitude towards morality may nevertheless feel that, in some circumstances, the criminal law is an unsuitable supporter or enforcer of their opinions.

Immoral Minority?
A Liberal View

In supposed contrast to a religious fanatic's supposed view of law
and morality we will now turn to the supposed position adopted by
a trendy, liberal, *Guardian*-reading, Hampstead-living person. If
mythology has it that Catholics want to ban everything, then Hampstead (Wo)Man on the other hand is thought to want everyone to be
left to do their own thing. But of course Hampstead people have
many reasons for wanting the law to enforce morality. Hampstead
people will want laws against violence, theft and deception in order
to allow them to go about their daily business without fear of
physical attack or damage to their property.

As Hart has explained,[1] such laws are inevitable in any society
because human beings are mutually vulnerable, roughly equal in
their capacity to harm or help each other, of limited altruism and
living in a world of limited resources. If we are both selfish and yet
social animals, if we want to survive and if we live amidst only
limited resources then we must band together under the protection
of rules against murder and other violence and against theft. But just
because we all agree on these values it should not be thought that
they are somehow morally neutral, or amoral. A law on theft pre-
supposes a system of property, for example, but the capitalist value
of private property might be considered immoral by those who see
the force of Proudhon's aphorism that 'all property is theft'.

Moving beyond basic survival, the whole edifice of liberal society
depends on the existence of law.

The essential point to remember about liberal attitudes to law
reform is that liberals, like everyone else, want the law to enforce
morality — *their* morality of liberalism. To say that the law should
not condemn homosexuality, for example, is perhaps to say that the
law should respect citizens' autonomy over their own sexuality. But
autonomy is a moral value just as much as the belief that homosexuality is unnatural is a moral value. Of course it is a more attractive
value to liberals, otherwise they would not count as liberals, but
there is no cause to regard liberalism as necessarily a superior creed
solely because it is sometimes represented as being morally neutral.

It is not neutral. It is partisan, affirming the value of freedom or of autonomy or liberty. That is one vision of morality and one which many of us find attractive, but it needs to be judged on its merits. Leaving people as free as possible to lead their own version of the good life may seem a praiseworthy approach. But if an Islamic fundamentalist is convinced that he knows what is the good life for us all, then he is entitled to put up for our consideration the view that we should only be 'free' to follow his vision of the good life. We then have a moral choice between one creed (liberalism) and another (Islam).

I have just argued that liberals want the law to reflect their morality of liberalism. But what does liberalism entail? 'Liberty' would seem to be the shortest and most appropriate answer. What is liberty? Is it freedom in the sense of an absence of restraint? Or does genuine freedom involve the provision of opportunities and resources so as to control one's life effectively? Is the Ethiopian famine victim, left alone in the desert, 'free'? Or is he only free when outsiders 'interfere' and provide him with food and the infrastructure for a self-sufficient future?

Perhaps the term 'autonomy' is more useful than the wider concept of freedom or liberty. Autonomy[2] is the capacity to think and to act on one's reasoning, to determine the course of one's life by oneself. Autonomy is a capacity which may be achieved to a greater or lesser degree. It is not necessarily achieved by leaving each other completely alone. We can help others attain an autonomous life by educating them, for example, thus developing their critical faculties of reasoning. We can also help others by providing them with options from which to choose their preferred life-style. The famine victim or, less dramatically, the long-term unemployed can be helped towards an autonomous life by the provision of resources and opportunities through public or private agencies.

Hence liberalism, understood in this light, does not mean that the government must always refrain from action and thus leave citizens 'free'. On the contrary, it might have to take positive steps in order to create the conditions under which citizens can act autonomously.

We are beginning to resolve an apparent paradox. Those who minimise state interference with the economy are often those who seek to maximise state regulation of sexual behaviour. And those who do not wish the law to enforce sexual morality are often keen for the law to insist on high standards of commercial morality, thereby protecting the consumer. Are those attitudes internally inconsistent?

Liberalism based on a respect for autonomy can explain a reluctance to enforce sexual morality resting alongside a willingness to enforce commercial morality. In both cases, of course, the liberal might want to protect people from coercion. Coercion might involve the subordination of another's will, whereas autonomy involves an exercise of reason. A dislike of coercion and a respect for autonomy therefore justify laws against coerced sex (rape, sexual offences with children) and laws against coerced contracts (bargains struck under duress).

But where coercion is not suspected, the liberal will generally wish that the law does not regulate sexual morality. The reason is that this respects others' autonomy.

In the commercial sphere, however, the individual may not have the wherewithal to act autonomously. He needs help to operate in a society which genuinely provides him with options from which to choose autonomously his lifestyle. He needs a job, money and adequate leisure in which to live his vision of the good life. In order to provide these conditions, the liberal may well want a whole range of state action, much of it flowing from the law.

It seems, therefore, that the idea of the liberal as an anarchist is another misconception. In order to create a liberal society, the law will have to create the right conditions for autonomous lives. The law will have to enforce a morality, albeit the morality of liberalism.

In practice, of course, one person's liberty can infringe another's freedom of action, so the liberal will at most support the maximum amount of individual liberty compatible with a like liberty for all. As regards law and morals, therefore, I do not accept a deep-rooted controversy between liberals and Catholics. Indeed it should be possible to think of liberal Catholics or, depending on the circles in which they move, Catholic liberals.

Amoral Law?
A Lawyer's View

Although law students are not always led to think of their traditional subjects in this way, even the most basic elements of a law course (and even those which have nothing whatsoever to do with sex) are inextricably bound up with morality or justice. Behind the technical law of tort or contract, for example, there are competing moral principles at stake.

Although the subjects may not always be presented in this light, the most basic aspects of civil law rest squarely on a moral code. The law of tort and particularly the law of negligence is, at least superficially, bound up with the moral principle that you should not harm your neighbour and that if you do you should compensate him. As Lord Atkin put it in the leading case of Donoghue v. Stevenson 'the liability for negligence is no doubt based upon a general public sentiment of moral wrong-doing for which the offender must pay'.[1] Lord Atkin went on to state the law of negligence by adapting the ethos behind the parable of the Good Samaritan: 'the rule that you are to love your neighbour becomes, in law, you must not injure your neighbour'.

As for the law of contract, its reliance on morality is well expressed by Professor Atiyah:

The law reflects to a considerable extent the moral standards of the community in which it operates. This has been especially true of England where moral standards have been, until very recently, almost identical with Christian standards and where the persons responsible for the development of the law have, almost without exception, been devout Christians. It is therefore not surprising to find that behind a great deal of the law of contract there lies the simple moral principle that a person should fulfil his promises and abide by his agreements.[2]

It is, as we have seen, often those who regard themselves as liberal who are most concerned for the law to *intervene* in contracts to enforce morality by countering inequality of bargaining power.

Consumer legislation is designed to protect the unwary consumer from hard-nosed businessmen who may otherwise extract 'unfair' deals. The law of contract can be seen as a tug-of-war between two moral principles. On the one hand we should keep our promise. On the other hand, we should not exploit people's economic weakness or their gullibility in getting them to promise something on very disadvantageous terms.

Land law is a third central element in legal studies. Its technicalities can sometimes seem far removed from great questions of morality. But the whole notion of property and ownership is full of moral import. This may become clearer if we think of an example in which respect for private property is *not* upheld by the law. Squatting, or occupying vacant property without the owner's consent, can in the course of time lead to the squatters acquiring some rights over the property. Is that moral? Is it moral, on the other hand, to allow property to remain vacant when there are homeless families? The renting of accommodation produces various moral dilemmas for land lawyers to solve. In particular, the Rent Acts seek to protect tenants from unfair rents. This is an attempt to resolve a classic conflict of moral values. Should we place more emphasis on the liberty of the landlord to profit from his own property or on the 'right' of everyone to have a roof over his head at a 'fair' rent? Are we for liberty or equality? Are we in favour of freedom or paternalism? Perhaps these are false antitheses, at least in this example. But it should be clear that the Rent Acts rest on a moral judgement that the law should intervene to enforce a particular conception of justice.

The Rent Acts also illustrate another feature of disputes over law and morals. Whatever our resolution of the moral rights and wrongs of charging 'excessive' rents, will the Rent Acts actually help the disadvantaged tenants? Apart from making value judgements, we have to make predictions as to what will happen under alternative legal schemes. If, with the best intentions in the world, we insist on 'fair rents', this may rebound to the detriment of poor tenants. Landlords may feel that it is not worth their while to use their property for rented accommodation because the return is too small. This would decrease the stock of available private rented accommodation which would leave more people unable to support themselves. It would also push up the level of the 'fair rent' for others if that rent bears any relation to market forces. We are here touching on economics, so this may be an appropriate time to observe that underlying the moral principles which seem to explain tort, contract and the rest of our legal system, there may well be an economic rationale.

Thus tort is seen these days as a question of loss- or risk-distribution more than as an application of the principles of fault. Fault is often the conceptual key which unlocks the insurance funds so that the cost is distributed amongst the payers of premiums. The law may be simply reflecting good economic sense.

Similarly, in contract it may be naïve to believe that promising is of much significance. In the modern world, we are constantly making contracts without understanding all the small print. If a contract is broken it is exceptionally rare for a court to order its fulfilment (known to lawyers as specific performance). Some cash compromise is worked out. Thus contract can be seen as revolving around the factors of benefit, detriment and reliance.

But even if economics could satisfactorily account for the law, that does not show that the law is amoral or morally neutral. For economics itself rests on moral values. Adam Smith, the economist, for example, took great care to stress that the 'invisible hand' of the free market can guide us through the rational pursuit of our own self-interest, towards an outcome which benefits society as a whole. Beneath the surface of economic arguments, moral disputes are struggling for air, often disguised as definitions of what counts as a 'cost' or a 'benefit'.

So the law student is not simply studying a morally neutral system of rules. When the rules are unclear or run out, as often happens in the topics covered at university, the moral nature of choices is often easier to discern. But even the settled rules have their roots in moral values. The law student may seek an economic route of explanation away from the moral dilemmas but that is only an intermediate step. At bottom, the legal system *and* our economic analyses reflect choices of what is good and bad, right and wrong, of what ought to be done and ought not to be done. However these are dressed up, the underlying moral tensions remain.

One final, brief example may help to show the moral problems lurking beneath the legal rules. Imagine a law student in 1984 learning 'the rules' concerning the powers and responsibilities of a local authority and its social workers with respect to children who may be at some risk of abuse in their family home. The future City solicitor may be unlikely to meet this problem in practice and fortunate enough to have had no personal experience of it. The rules may seem uncontroversial, even boring.

But in 1985 the topic suddenly became headline news. Case after case of child abuse appeared in the newspapers and on television, culminating in the trial over the death at the hands of her father of Jasmine Beckford, a little girl under the care of her local authority. An inquiry by a senior barrister, Louis Blom-Cooper QC, revealed many problems with the implementation of the law (although few with the content of the law itself).

The whole issue of child abuse, children's rights, parents' rights and social workers' duties came out into the open. The most bored students could not fail to be interested in how the law, and social workers' practice, should resolve the tension between respect for the privacy and primacy of the family on the one hand, and concern for the well-being of a child in an unstable family on the other hand. Today's technical rule of law can easily become tomorrow's tragedy or tomorrow's confrontation between privacy and safety, the family and the state.

The law is not, cannot be, and should not be, morally neutral. But *which* moral values should it seek to enforce?

5

Mill's 'Harm-to-others' Principle

John Stuart Mill's statement of principle reads as follows:

The only purpose for which power can be rightfully exercised over any member of a civilised community against his will is to prevent harm to others. His own good, either physical or moral, is not a sufficient warrant.[1]

Mill's harm-to-others principle seems simple, but it gives rise to many problems. Firstly, what is harm? Secondly, who counts as others? Thirdly, given that Mill says that the prevention of harm to others *can* be the reason for restrictions, he presumably thinks that there is still a question as to whether we *should* restrain conduct even if it causes harm to others. The only reason for which power *can* be exercised over citizens against their will is to prevent harm to others, but Mill does not say power *must* be used to prevent harm to others. Harm to others is a necessary but not a sufficient condition. We may have to calculate the harm caused by intervention and weigh it against the harm caused by the initial action. A fourth issue is, why should we accept the harm-to-others principle as the exclusive justi- fication for intervention? Why not allow society to be paternalistic, preventing people from harming themselves?

Mill clearly intended harm to include physical harm, but what about mental, moral, emotional and spiritual harm? The criminal law's control over pornography, for instance, would depend partly on whether one counts the offence felt by non-participants as harm or whether one counts any moral harm as sufficient reason for intervention. Or, to take another example, adultery does not physi- cally harm participants, but surely it does cause emotional harm to others. But once Mill's principle is enlarged to include non-physical harm it can be extended to the point at which it allows intervention in practically all circumstances.

Business life, for instance, is all about competing and therefore, in a sense, 'harming' competitors by succeeding. The morality of the business and financial worlds is itself a topic which merits serious study. Samuel Brittan, the economic commentator, has given us some food for thought on the morality of the market and hence on what should count as 'harm' in these circumstances:

The shooting of one's competitors is not an acceptable way of maximising profits or even minimising losses. Nor is the bribery of legislators. But there are less banal and obvious questions. Is it legitimate to try to monopolise a market in the absence of strong restriction practices legislation? If there are such laws, should we co-operate actively, or merely conform? If certain markets, such as those for labour, are clearly malfunctioning, is an employer who tries as a deliberate policy objective to provide more jobs a social benefactor? Or is he meddling in matters outside his influence, and would society be better off if companies confined themselves to promoting share-holders' financial interests?[2]

Who counts as 'others' is obviously at the centre of disputes over abortion and experimenting on embryos. There is no doubt that abortion causes physical harm to the foetus. There is no doubt that experiments harm the embryo. But the debate over these issues concerns the question whether the embryo or foetus is a person or potential person who deserves society's protection. Just as the harm-to-others principle does not help us define harm, so it does not help us to define others since we can see that many important legislative tussles hinge precisely on what is harm and who are others. Bandying Mill's phrase around is far from conclusive.

Our next doubt about the efficacy of Mill's approach is that even if we establish harm-to-others, much work remains to be done. Many debates over law and morality are not about whether or not harm-to-others is caused, but revolve around whether the harm is sufficiently serious for the State to intervene and whether the State's intervention will cause more problems than it solves.

Our final problem is that paternalism (preventing someone from 'harming' himself) is excluded by the harm-to-others principle. But why shouldn't society be concerned to protect individuals from their own folly? Whether it concern alcohol- or drug-addiction or under-age sexual activity, parents and friends *are* paternalistic in practice. Mill says that an individual 'cannot rightfully be compelled to do or act because it will be better for him to do so, because it will make him happier, because in the opinions of others it would be wise or even right'. Yet, in fact, these are precisely the reasons behind many parental and societal attitudes. Actually, Mill deems that 'these are good reasons for remonstrating with him or reasoning with him or persuading him or entreating him but not for compelling him'. But by now Mill is treading a very thin line. If we can be paternalistic in our conversation and attitudes, what is so different about being paternalistic through the law? To my mind, at least, it is little more than the question of the degree of effectiveness. The law may often be more powerful in coercing citizens. On the other hand, a parent's withering look may be more effective than the possibility of a small fine imposed by the State.

In fact, Mill himself offers some exceptions where society can intervene to promote an individual's own good even where there is no harm-to-others. One exception is 'backward states' where people are not 'capable of being improved by free and equal discussion'. Another is where there is ignorance of danger: in particular, paternalism is allowed with regard to children. Here again the exceptions, once their rationale is understood, could eat up the rule. Why can we be paternalistic over children? The answer is because children do not fully understand the consequences of their actions. But then, who does? One argument made against advertising by big business and distortion by the media is that these may colour our attitudes so that we act out of ignorance, an ignorance often fostered by those in power. Even less radical views will concede some limitations to our foresight. Certainly, the case of heroin becomes intriguing at this point. Mill would allow us to prohibit heroin from being sold to children but the harm-to-others principle might well allow it to be used by an adult. Yet the point of Mill's whole approach is to create a climate in which we can make autonomous decisions. Mill wants us to be able to run our own lives, to choose our own vision of the good life or, in modern parlance, to do our own thing. A child will never be able to make fully free adult decisions if he stumbles into drug-addition at an age when he could not appreciate the consequences of his actions. But this applies also to the adult who becomes a drug addict.

If we regard freedom or liberty or autonomy as something existing over time, and if Mill's aim is to increase our autonomy overall, it might well be the case that a short-term restriction of liberty enlarges or preserves our long-term freedom. If we stop even the adult from taking heroin once, we have infringed his autonomy. Nevertheless, we have stopped him from taking the first step to drug addiction which would end up with his inability to make *any* autonomous decision. Some would argue that once heroin has been taken by an individual he can no longer genuinely choose whether or not to continue taking it. In that context the law's paternalism in prohibiting the sale or use of heroin can be seen as liberty-enhancing in the long run at the same time as being liberty-restricting in the short term.

Our last criticism, then, of the harm-to-others principle is that it is counter-intuitive. We often think paternalistically, we are often right to think paternalistically, even when judged by liberal standards, and if it is acceptable to remonstrate or persuade, it may well be acceptable to use a more effective form of paternalism, namely the law.

Merely incanting 'harm-to-others' is not sufficient to provide a recipe for when the law should enforce morality. At best, it provides a starting-point. At worst, it begs all the important questions.

6

The Hart–Devlin Debate

Lord Devlin and Professor Hart have developed the debate about Mill's harm-to-others principle in the course of their own controversy over law and morals.[1]

The 1957 Report of the Wolfenden Committee on Homosexual Offences and Prostitution recommended the decriminalisation of homosexual acts between consenting adults in private. Its justification was

the importance which society and the law ought to give to individual freedom of choice and action in matters of private morality. Unless a deliberate attempt is to be made by society, acting through the agency of the law, to equate the sphere of crime with that of sin, there must remain a realm of private morality which is, in brief and crude terms, not the law's business.

The Wolfenden Committee endorsed Mill's statement and put forward its own similar principle. The committee said that the function of the criminal law

is to preserve public order and decency to protect the citizens from what is offensive or injurious and to provide sufficient safeguards against exploitation and aggravation of others, particularly those who are specially vulnerable because they are young, weak in body or mind, inexperienced, or in a state of special physical, official or economic dependence. It is not, in our view, the function of the law to intervene in the private lives of citizens.

When he first read the Wolfenden Report's insistence on a 'realm of private morality', Lord Devlin agreed with it. Indeed Sir Patrick Devlin, as he then was, had given evidence as a High Court judge to the Wolfenden Committee and had himself recommended an easing of the restrictions on homosexuals. In preparing a distinguished lecture on jurisprudence, however, he began to have second thoughts. Although he was still in favour of Wolfenden's recommendation for law reform in the case of homosexuality, he began to doubt the wisdom of the Wolfenden approach as a general guide to the legal enforcement of morals.

Lord Devlin posed three questions:

1. Has society the right to pass judgement at all on matters of

morals? Ought there, in other words, to be a public morality or are morals always a matter for private judgement?

2. If society has the right to pass judgement has it also the right to use the weapon of the law to enforce it?

3. If so, ought it to use that weapon in all cases or only in some; and if only in some, on what principles should it distinguish?

Lord Devlin's answers may be summarised as:

1. Society does have the right to pass judgement on morals. What makes a number of individuals into a 'society' is precisely a 'shared morality': 'If men and women try to create a society in which there is no fundamental agreement about good and evil they will fail; if, having based it on common agreement, the agreement goes, the society will disintegrate'.

2. Society does have the right to use the law to enforce morality 'in the same way as it uses it to safeguard anything else that is essential to its existence'. Thus, 'The suppression of vice is as much the law's business as the suppression of subversive activities'.

3. But society should only use the law in some cases.

Lord Devlin suggests four guidelines, all of which are principles of *restraint* in the way society should use the law to enforce morals:

(i) 'Nothing should be punished by the law that does not lie beyond the limits of tolerance'. That tolerance should extend to the maximum individual freedom consistent with the integrity of society. The limits of tolerance are reached at a 'real feeling of revulsion', not merely at 'dislike' of a practice.

(ii) '(T)he extent to which society will tolerate — I mean tolerate, not approve — departures from moral standards varies from generation to generation'.

(iii) '(A)s far as possible privacy should be respected'.

(iv) 'The law is concerned with a minimum and not with a maximum standard of behaviour'.

Professor Hart prefers to see Lord Devlin's questions collapsed into this formulation: 'Is the fact that certain conduct is by common standards immoral sufficient to justify making that conduct punishable by law? Is it morally permissible to enforce morality as such?'

Lord Devlin is taken to answer 'yes' whereas Professor Hart says 'no'. Hart adopts a different starting-point. He develops Mill's harm-to-others principle so as to include physical harm to oneself as also constituting a ground for intervention.

This can be characterised as a paternalistic approach. Hart does not want to include *moral* harm to oneself as a justification because that would lead him perilously close to Devlin's position, which Hart dubs 'legal moralism' (the prevention of conduct because it is deemed to be inherently immoral, whether or not anyone is 'harmed'). Much the same would then happen under both theories but for different reasons: Hart's focus is on the individual whereas Devlin's focus is the society. But indeed it may well be that both protagonists *are* fairly close to one another's views, with the major dispute turning on sexual morality.

Where does Hart think Lord Devlin falls into error? According to Hart, Devlin's mistake lies in his

undiscussed assumption. This is that all morality — sexual morality together with the morality that forbids acts injurious to others such as killing, stealing and dishonesty — forms a single seamless web, so that those who deviate from any part are likely or perhaps bound to deviate from the whole. It is of course clear (and one of the oldest insights of political theory) that society could not exist without a morality which mirrored and supplemented the law's proscription of conduct injurious to others. But there is again no evidence to support, and much to refute, the theory that those who deviate from conventional sexual morality are in other ways hostile to society.

Professor Hart's critique of Lord Devlin became the orthodoxy of the 1960s but, nowadays, Lord Devlin's approach seems to be coming back into vogue. Lord Devlin's answers to his own three questions seem eminently plausible. Yes, society has the right to pass judgement on morals. Yes, it has the right to use law to implement such judgements. No, the full panoply of the law should be reserved for particular dangers. Some principles have been given by Devlin, and very helpful ones at that.

Yet Professor Hart's approach *also* has its merits. He is probably right to observe that sexual morality is less settled and immutable than other areas of morality, at least in our culture. Hart's emphasis on 'critical' morality rather than 'positive' morality is also an aid to clear thinking. If we are tempted towards legal moralism, for example, we should at least clarify whether we are arguing for the law to enforce the morality which happens to be accepted and shared by society ('positive' morality) or to enforce what we regard as an ideal morality (a 'critical' approach which might be based on rights or utility). Otherwise, there is a danger of entrenching society's prejudices under the banner of morality. Society, or at least its leaders, has accepted all manner of views which we would now criticise. Does legal moralism simply justify the status quo of, say, Nazi

Germany or of South Africa's system of apartheid? Should we not instead be criticising such 'positive' immorality?

Hart is careful to spell out exactly why we should look before we leap from moral censure to legal enforcement. He gives us four grounds for caution:

(1) 'the actual punishment of the offender', because depriving someone of their freedom of movement (through imprisonment) or of their money (through a fine) clearly harms them.

(2) the 'unimpeded exercise by individuals of free choice may be held a value in itself with which it is prima facie wrong to interfere'.

(3) the unimpeded exercise of free choice 'may be thought valuable because it enables individuals to experiment — even with living — and to discover things valuable both to themselves and to others'.

(4) as far as sexual morality is concerned, 'the suppression of sexual impulses' is generally 'something which affects the development or balance of the individual's emotional life, happiness and personality'.

The idea that freedom is valuable in itself is disputed by Devlin and many others. Devlin is less easily persuaded than is Hart of the value of experiments with life-styles. But Hart's list can be regarded as a partial justification for restraint in using the law to enforce morality and thus as an explanation of why we should adopt principles like Devlin's restraining ideas of tolerance and privacy.

Finally, Hart makes us wary of legal intervention based solely on the distress caused to non-participants by the mere knowledge that others are behaving in a certain way. If we are offended by *witnessing* some act of which we disapprove, that might be appropriately termed 'harm'. But distress at the mere *knowledge* that others are acting in ways of which we disapprove would not come within Hart's definition of 'harm'. He claims that 'Recognition of individual liberty as a value involves, as a minimum, acceptance of the principle that the individual may do what he wants, even if others are distressed when they learn what it is that he does unless, of course, there are other good grounds for forbidding it'. Even those who disagree or think that such distress is a signal that there *are* 'other good grounds', will appreciate that Hart is once again warning us about moving automatically from moral disapproval to legal action.

Our conclusion may well be, of course, that we do not have to choose between the great liberal judge and the great liberal professor. Both offer insights which contribute to our understanding of a complex problem. If one is right, it does not necessarily mean that the other is wrong.

A phrase sometimes used in jurisprudential discussions is apposite here. Devlin and Hart each present 'one view of the cathedral'. The useful, if pretentious, allusion is to Monet's twenty studies of the cathedral at Rouens. Monet painted it in different lights at different times of the day. Hart and Devlin are painting a picture of law and morals from different vantage points. Both can deepen our understanding of law and morals. Both reveal something of their own vantage point from the picture which they paint.

So my contribution to the discussion of law and morals is not put forward in a hostile spirit. Instead, this book seeks to make known and build upon the earlier works. But I do think that a clearer view of the cathedral can be attained if we look at it in a different light or from a different perspective.

7

A Different Framework

I have sketched some problems with Mill's harm-to-others principle and some reasons for doubting whether Devlin or Hart or anyone else has provided a complete solution. This is not surprising.

A Different Approach

There is no need to wait for a generalised statement of principle with which to analyse moral problems and the law's response thereto. If an agreed and helpful standard existed, it might enable us to achieve consistency across time and subject matter. But as there isn't an agreed and helpful standard, we would only be concealing important disagreements over society's values by trying to use some pithy formula. Instead, we need to rethink and reargue the old problems in their new contexts. There is a danger in spending so much time on what Mill, Hart and Devlin thought in general, that we never get round to deciding what *we* think on *particular* topics. I have accordingly taken the risk of failing to do justice to the richness of their great ideas in the belief that this book may interest readers sufficiently to stimulate independent study of the classic theorists.

But I would like to offer a different framework which has evolved through discussing law and morals with innumerable students. Whereas the harm-to-others principle tends, in my experience, to obscure the real sources of disagreement in controversies, I have found that the separation of factual predictions and moral assumptions helps to clarify thought.

Discussions on law and morals benefit from the attempt to locate precisely the points of dispute. These tend to be, firstly, different predictions of the consequences which would be likely to flow from the alternative courses of action proposed for the law; and secondly, different sets of moral values, or different conceptions of those values, or different weights attached to the values.

Recognition of these elements does not provide an answer to the question of what the law should do about a particular moral dilemma. But such recognition might help us to structure our own thoughts and society's decision-making procedures in an appropriate way to resolve the dispute.

An example might be the issue of violence on television and whether the law should impose censorship. Our first concern might be to argue about the consequences of allowing violence on television. Some of us might be prepared to restrict freedom of expression if a direct link to violence in the real world could be established, but might dispute that any such connection has been demonstrated. Whether or not we can reach agreement on the effect of violence on television, we then have to establish the significance we attach to broadcasters' freedom of expression as against the familiarity of our society with violence.

My emphasis on these two aspects of controversies is intended to direct us towards concentrating on the same points and explicitly on the precise issues of disagreement. Once we establish what is at stake, we can go away to research into the effects of violence on television or to think about the value of freedom of expression before returning to the fray.

Within the second area of competing moral outlooks, it is not my intention to suggest a way of resolving all controversies; but we can perhaps post some guidelines.

First, some values will usually feature in the equation on the side of restraint, namely Devlin's 'elastic' principles such as tolerance of individual liberty and respect for privacy.

Second, it should be possible to trace the values on which we are relying back to some test of morality, such as the premise of utility or intuitions as to rights. These alternative bases of moralities are not themselves susceptible to 'proof' (otherwise they would not be bases, nor would all of them still be competing for our attention). Nevertheless, there is a virtue both in rationality and in appreciating its limits.

Third, if we find ourselves regularly regarding certain values or interests as especially worthy of strong protection, we might call these interests 'rights'.

A right exists when 'a person has a guaranteed expectation that some choice of his will be respected or some interest protected or advanced.'[1] Rights, in one vision of morality, are paramount considerations. Rights trump other moral values. In particular, rights will normally trump arguments based on utility or the collective good of society. They protect the individual from being sacrificed in the interests of society as a whole.

But why not sacrifice an individual for the good of the rest of us?

The assertion that there is a right to life or to freedom of expression, may be one moral conclusion. Yet another view might be that all such claims can be defeated by the requirement of society.

Time and again, then, the debates which follow could be represented as controversy between those who believe in rights and those who believe in utilitarianism.

Believing in rights does not solve all one's moral problems. What are the specific rights and what do they require in particular circumstances? Moreover, rights may conflict with one another and a balance will have to be struck.

Nor does believing in utilitarianism make moral dilemmas easily resolvable. Utilitarianism requires weighing up the happiness or pleasure caused by an action or a rule, and weighing that against the unhappiness or pain caused by it. But how can we measure happiness or pleasure, how can we compare your pleasure with my pain and how can we know what pain or pleasure will result?

It is sometimes said that a rights-based approach may be all very well for one's private morality but that utilitarianism is the inevitable criterion for a public morality. Public morality has to involve compromise, give and take. It cannot deal in absolutes, especially where the law is concerned. I do not find this supposed distinction between 'private' and 'public' particularly helpful in relation to morality. Certainly, I see no reason to assume that the law should always plump for utilitarianism.

Nevertheless, there is here perhaps a warning which is of vital importance to our endeavours. Where we are trying to influence the actions of others, as through the law, we should always remember that there are certain 'rights' (for example, to privacy, liberty) or certain 'pains' (for example, the unhappiness caused by coercion restricting one's liberty) which militate against intervention. As I have tried to indicate by the alternative formulations of 'rights' and 'pains', this can be accommodated within a rights-based or a utilitarian moral code. It follows that one should never leap from the premise that something is immoral to the conclusion that it should therefore automatically be made illegal.

Whether or not to make the leap is itself a moral problem and one which requires careful thought. Given that we have different moral perspectives, let alone different perceptions of the factual context and consequencs, we should expect disagreement on the morality of the relationship between law and morals.

I hope that the injunction to focus on predicted consequences and moral values will serve as a useful way in which to clarify our thoughts. Occasionally, I will also use the language of the traditional harm-to-others test. But there are no easy answers. It is not the aim of this book to determine the reader's moral outlook. But a consideration of the following controversies should help identify some of the recurring values at stake and the implications of assigning different weights to those values.

There is one other theme on which the reader should ponder. What should we do, in our democracy, when there is disagreement about the proper legal response to a moral dilemma? Should we count up all those in favour of, and all those against, a proposal, and implement the wishes of the majority? Or should we do what is 'right', whatever the majority thinks, and if so, of course, how do we determine what is right?

I will not refer to that issue expressly until the conclusion and even then I have no stunning answer. But I am concerned that we focus on the structures by which these moral dilemmas are in practice resolved. Sometimes we rely on backbench MPs' own research and intuitions. Sometimes we turn to a government-inspired committee of inquiry. On other occasions, the courts determine the outcome. In other contexts, administrators have the effective power. Is there any logic in all this?

As a final reminder before we tackle some pratical problems of law and morals, we ought to analyse our own methods of analysis.

In my view, it is most helpful to recognise that these controversies can be reduced to differences of opinion on the practical effects of various options and on the morality of the conduct in question.

But, traditionally, people have relied on Mill's harm-to-others principle. If you cannot resist adopting that approach, I would at least stress that the important questions to ask, if Mill's principle is accepted, are:

1. What is the harm?
2. Who or what is harmed?
3. How serious is that harm?
4. Are there any countervailing benefits?
5. Do those harmed need or deserve society's protection?
6. What level of protection is most appropriate, ranging from social pressure to life imprisonment for those causing the harm, bearing in mind the costs or disadvantages of such intervention?

Whichever structure one finds most useful can now form the framework for an examination of some controversies in law and morals. Moreover, as we reflect on each topic we should pause to examine the consistency of our moral stances. We shall begin by questioning the internal consistency of the Warnock Report. Then, in the context of the Gillick case, we should ask whether those who claim that parents have absolute rights over their 15-year-old children are thereby committed to accepting that parents have the right to sanction experiments on their 15-day-old embryos, or the right to abort their 15-week-old foetuses.

But it is also important for readers to test their attitudes across the range of topics we discuss — and beyond. In relation to the Warnock Report, for example, it is obviously interesting to compare our attitudes to experimenting on early embryos with our views on aborting more mature foetuses. Less obviously, perhaps, those who condemn surrogate motherhood might ask themselves whether they are thereby condemning the use of nannies and child-minders. They might reject the analogy, or soften their attitude to surrogacy, or harden their attitude towards the practice of post-natal surrogate parenthood which affects so many more people than does the use of ante-natal nannies. Or they could reject the ideal of consistency, arguing that we hold many conflicting values which we balance in different ways. But nobody should pretend that it is easy to resolve all these dilemmas.

As we consider each problem, we should bear in mind the wise words of the Archbishop of York: '. . . in practice most contentious ethical issues arise in the murky area where principles conflict, facts are ambiguous and differences are largely a question of degree'.[2]

Embryo Experiments and Surrogate Motherhood: Warnock

8

The 1984 Warnock Report recommended the establishment of an independent statutory body to monitor, regulate and license infertility services and embryo experiments. Experiments should be lawful, according to the Warnock majority, during the first fourteen days of development. Sperm-, egg- and embryo-donation should be facilitated in that resulting children should not be considered illegitimate, the non-contributing parent(s) might be lawfully registered as the parent(s) on the birth certificate, and donors should be relieved of parental rights and duties in law. But surrogacy arrangements were frowned on by the majority of the Committee. They recommended that the operation of surrogacy agencies should be criminally prohibited and that any private surrogacy arrangements between individuals should be illegal and unenforceable in the courts (although not criminal, since it would not help the child to be born into a family threatened by imprisonment).

Reliance on Utility or Feelings?

The Warnock Report fails to clarify its attitudes to theories of morality and its approach to the relationship between law and morality. These, of course, are our central concern.

On morality, the key recommendation rests on a utilitarian approach even though the Report had already provided a damning indictment of utilitarianism. Thus, research on embryos is justified on utilitarian grounds by a majority of the Committee:

We do not want to see a situation in which human embryos are frivolously or unnecessarily used in research but we are bound to take account of the fact that the advances in the treatment of infertility, which we have discussed in the earlier part of this report, could not have taken place without such research; and that continued research is essential, if advances in treatment and medical knowledge are to continue. A majority of us therefore agreed that research on human embryos should continue.

Yet the flaw in this reliance on the benefits of research was ably explained in the Report's own Foreword:

A strict utilitarian would suppose that, given certain procedures, it would be possible to calculate their benefits and their costs. Future advantages, therapeutic or scientific, should be weighed against present and future harm. However, even if such a calculation were possible, it could not provide a final or verifiable answer to the question whether it is right that such procedures should be carried out. There would still remain the possibility that they were unacceptable, whatever their long-term benefits were supposed to be. Moral questions, such as those with which we have been concerned are, by definition, questions that involve not only a calculation of consequences, but also strong sentiments with regard to the nature of the proposed activities themselves.

As far as the relationship between morality and law is concerned, the Report must again be dubbed inconsistent. The acceptance of artificial insemination by a donor should be contrasted with the rejection of surrogate motherhood. Those who object to sperm- (and, elsewhere, egg- or embryo-) donation are told that they should not impose their moral standpoint on those who disagree. But a majority of the Committee is prepared to ban surrogacy agencies when the same point could be made.

The Report does not explain how to bridge the gap between deciding that something is immoral and deciding that it should be illegal. Before we examine that in some detail, it is instructive to get a flavour of the kinds of issues to be decided. What I call the 'Hamsterman'[1] test may prove a salutary experience for those who regard themselves as 'liberals' on the law and morality question.

Should we allow the development of 'Hamstermen' or, more probably, 'Gorillamen'? In my experience, hitherto unshockable liberal students have unanimously drawn the line at this point of discussing the Warnock Report. I have yet to meet any person (or indeed any hamster or gorilla) who has expressed acceptance of such developments beyond the two-cell stage. But why do we object to trans-species fertilisation involving humans? Are farming techniques which cross-fertilise non-human species also unacceptable? Are we speciecists and, if so, is there anything wrong in such an attitude? We seem to share an intuition to protect the boundaries of our species or, as it is often expressed, to preserve the Dignity of Humanity. Our response is one of disgust and that itself is significant. If forced to rationalise this emotional reaction, sometimes dubbed the 'yuk' factor, we may flounder but that does not necessarily invalidate the intuition.

What about a ban on the placing of a human embryo in the uterus of another species for gestation? While the Report thinks of this practice as only a 'possibility', other evidence suggests that it is already happening. Who wants camels as surrogate mothers? Again, I suspect many regard Warnock's recommendation against trans-species gestation as correctly reflecting a general repugnance at such an idea. On the other hand, Warnock does not *argue* for this view. Is the statement of moral intuition enough? On both trans-species fertilisation and gestation, I suspect most people regard argument as unnecessary. That may be because the argument is obvious, although not to everyone, but it may be because we have reached the limits of rationality's usefulness.

Another oddity is that at one point the Committee rejects the benefits of the research argument in deference to feelings, whereas its major recommendation would seem to imply the contrary:

It has been suggested that human embryos could be used to test the effects of new developed drugs or other substances that may possibly be toxic or cause abnormalities. This is an area that causes deep concern because of the possibility of mass production of *in-vitro* embryos, perhaps on a commercial basis for these purposes. We feel very strongly that the routine testing of drugs on human embryos is not an acceptable area of research *because this would require the manufacture of large numbers of embryos. We concluded however that there may be very particular circumstances where the testing of such substances on a very small scale may be justifiable* (my italics).

On the majority's earlier, utilitarian reasoning, one might have thought that the more experiments the better. If we accept Warnock's very strong feeling against the routine testing of drugs on human embryos, we should ask whether this means that feelings are sufficient warrant for action in themselves and whether such feelings are inconsistent with allowing *any* experiments on embryos.

Let us now turn to the issues which have dominated media reaction to Warnock: embryo experimentation and surrogate motherhood.

Surrogate Motherhood

The central paradox in the Warnock Committee's attitude to law and morality comes in its proposals on surrogate motherhood. Those who object to *in-vitro* fertilisation, artificial insemination by a donor and other forms of donation are told, in effect, 'if you don't like it don't do it but don't force your views on those who wish to avail themselves of these opportunities'. Yet those who object to surrogacy *are* allowed to impose their views on others. Why? The reasons are far from clear. At one point it looks as if public hostility, or perceived public hostility as whipped up by the popular press seems to be the key factor. At other points it seems to be based on either possible psychological harm to the child or the surrogate mother or possible harm to the whole concept of the family and particularly motherhood. So it is disappointing that the Committee never gets round to analysing its own attitudes to enforcing morality through law.

In her introduction to the new edition of the report, Mary Warnock now belatedly acknowledges the Hart–Devlin debate. At the time of the Report, however, the Committee does not seem to have thought things through. For example, whether or not harm to the institution of the family should count as a reason for restricting or banning surrogate motherhood is vital. Can we extend Mill's harm-to-others principle to include harm to social institutions? The Catholic bishops' submission to the Warnock Committee thought that harm to social institutions was a legitimate reason for State intervention. The bishops' evidence constantly stresses the harmful effects to marriage and the family caused by sperm- or embryo-donation or womb-leasing. The bishops, in fact, develop the harm principle when they conclude that 'unwisdom which harms another, *or undermines a fundamental and valuable form of social life is the law's proper concern*' (my italics).

This makes sense if you believe that the individual needs both respect for his autonomy and the opportunity to live in a community in order to flourish fully as a human being. It may well follow from Hart's characterisation of us as both selfish and social beings that we need not only the opportunity to operate as an individual but also social institutions such as marriage and the family. If so, surrogate motherhood may well pose a problem, though not necessarily a serious one. But for all the high principles expended in discussing surrogate motherhood, in reality what has counted with Parliament has been media pressure and the practical difficulties of drawing any fine lines between permissible and non-permissible forms of surrogate motherhood. Hence the law's response has been swift and certain: commercial agencies for surrogate motherhood have been outlawed and in due course a fuller range of legal responses will follow.

The surrogacy issue is complex because it is difficult to predict accurately the consequences for all concerned. If we *knew* exactly how resulting children, surrogate mothers, commissioning couples and others involved would feel about surrogacy (both during the pregnancy and before and afterwards until the end of their lives), we would be in a better position to judge whether values of human dignity, parenthood and the family really were under threat. Another complication is the question whether it is any of our business at all. But while I think that our moral matrix includes respect for such values as privacy and tolerance, so that we are initially reluctant to act on external preferences (here, our dislike of surrogate motherhood), I would not exclude them from consideration. Here I side with the bishops because we *are* affected by what others do. Even Mill acknowledged this when he admitted that, 'the mischief which a person does to himself may seriously affect, both through their sympathies and their interests, those nearly connected with him and, in a minor degree, society at large'. Although surrogacy is life-giving, many feel that it is fundamentally flawed because, as Warnock concluded, 'That people should treat others as a means to their own ends, however desirable the consequences, must always be open to moral objection'.

Embryo Experiments

But why not apply that framework to the other, more important, issue of embryo experimentation? Instead, the fragile (9–7) Warnock majority switches to pragmatic or utilitarian arguments:

We are bound to take account of the fact that the advances in the treatment of infertility . . . could not have taken place without such research; and that continued research is essential, if advances in treatment and medical knowledge are to continue. A majority of us therefore agreed that research on human embryos should continue.

Despite their agreement that 'the embryo of the human species ought to have a special status' and that 'the status of the embryo is a matter of fundamental principle which should be enshrined in legislation' the majority therefore recommended 'that research may be carried out on any embryo resulting from *in-vitro* fertilisation, whatever its provenance, up to the end of the fourteenth day after fertilisation'.

Three members dissented in favour of a total ban on experimentation, while four dissented in opposition to research on embryos deliberately created for such purposes. Of course, we have already seen that the *whole* Committee had earlier realised the inadequacies of utilitarianism in its Foreword. Even if the majority has accurately weighed up the consequences, that is not the only basis for deciding the issue: 'Moral questions . . . are, by definition, questions that involve not only a calculation of consequences, but also strong sentiments with regard to the nature of the proposed activities themselves'. No critic could match the eloquence with which the Warnock majority once again condemns itself.

So although Warnock tried to dodge the fundamental issue of when human life or personhood beings, this is unsatisfactory. 'Instead of trying to answer these questions directly we have . . . gone straight to the question of how it is right to treat the human embryo' (para. 11.9). But we need to know whether the embryo counts as an 'other' for the harm-to-others test or for Warnock's own touchstone which may be characterised as a 'people-treating-others-as-a-means-to-their-own-ends' test. This is the question which needs to be resolved before we can apply that fundamental value of respect for innocent human life. Is the embryo to be regarded as a person, or a human, or a *potential* human person?

I find the Warnock majority unconvincing on this. You will not be surprised to learn that the Warnock majority has undermined its own position once again. Even the majority conceded that '*once the process has begun*, there is no particular part of the developmental process that is more important than another' (my italics). So why not start protection once the process has begun at day one, rather than at day fifteen? Even if you do not accept the early embryo as a 'person', the minority argument of potentiality[2] might give you cause to doubt the recommendation of the majority — surely from the beginning the embryo has 'potential for development to a stage at which everyone would accord it the status of a human person' and is it not 'wrong to create something with the potential for becoming a human person and then deliberately to destroy it'?

Whether or not children born from surrogates and embryos have rights or deserve our protection ought to be the central concerns in the post-Warnock debate about law and morality. While we *do* value tolerance and privacy and the freedom of others to do what they believe in, there are limits which we normally impose on such freedom. The limit is at least where the interests of third parties are concerned. From one view, embryos are exactly the voiceless minority who need and deserve society's protection.

Secondary Issues

This is not the place to examine the Warnock Report in minute detail. We have seen some problems in its approach to morality and law in the context of the central concerns: embryo experimentation and surrogate motherhood. But it is often the 'peripheral' issues in a controversy which make us think about fundamental problems. Let us briefly highlight one or two other points of interest.

One criticism of the whole Warnock exercise might be that too much emphasis has been placed on high-technology responses to the tragedy of infertility and not enough attention has been paid to preventing infertility from occurring in the first place.

Notwithstanding the precise terms of reference, the Report seems to take as its focus infertility. Chapter 2 is headed 'Infertility: The Scope and Organisation of Alleviation of Infertility'. We are not offered a definition of infertility, nor are its causes explained, nor are treatments other than *in-vitro* fertilisation discussed. If the object of the Inquiry was to focus on infertility, these should surely have been three initial matters of vital significance.

The Catholic Bishops' Joint Committee on Bio-Ethical issues, in contrast, has noted,

the Report's striking silence about the causes of infertility, and its consequent failure to consider how a social policy might seek to reduce infertility by attending to its causes. *In-vitro* fertilisation (IVF), when used as a clinical technique, is largely designed to solve those problems of infertility which are caused by tubal occlusion. But the commonest causes of tubal occlusion (accounting for about 90% of cases) are previous abortion, the use of the IUD as a contraceptive device, and sexually transmitted diseases. To have pointed to these causes would have taken courage. It would not, however, have implied that all cases of infertility result from such avoidable causes. And what is at stake is a matter of fact, of truths which it is irresponsible for society to conceal from its vulnerable members. Public policy should not ignore these facts when determining the proper distribution of society's inevitably limited resources for health care.

'Are the bishops right?' and 'If so, so what?' are important questions which merit answers but which the Warnock enterprise fails even to ask. Society certainly should ask whether prevention is better than cure (or rather 'alleviation').

Another area of interest is the eligibility for treatment. The Report *does* ask who should be eligible for infertility treatment but we are not really given the reasons for the Committee's conclusions, which are: 'we believe that as a general rule it is better for children to be born into a two-parent family, with both father and mother' but 'we are not prepared to recommend that access to treatment should be based exclusively on the legal status of marriage'. So the question for us is whether only 'good' prospective parents should be eligible. If so, does 'good' mean married or a couple or heterosexual? The Report is tantalising in raising questions about the eligibility of single homosexuals but ultimately it fails to counter the argument that 'a single person, whether man or woman, can in certain circum-

stances provide a suitable environment for a child, since the existence of single adoptive parents is specifically provided for in the
Children Act 1975'. Clearly, artificial insemination by a donor and
surrogate motherhood provide opportunities for the single heterosexual person and perhaps more intriguingly the lesbian or homosexual couple to 'start a family' in a way which may challenge
traditional notions of the family.

Within heterosexual relationships, the Report's rejection of marriage as the dividing-line for eligibility will annoy some and please
others. But both groups might have expected more discussion of the
arguments about whether a child's best interests are served by being
born of or for a married couple. A serious question which the Report
raises but again fails to resolve adequately is whether a couple with a
previous conviction for child abuse should be ineligible for infertility
treatment. A final point in this context is the question of priorities in
a world of scarce resources. Should treatment be available to those
who earlier opted for infertility through sterilisation but (perhaps
after remarrying) have now changed their minds, or to those who
have become involuntarily infertile but who already have children?

Then there is the question, noted earlier, of guarantees as to a
donor's race:

As a matter of principle we do not wish to encourage the possibility of
prospective parents seeking donors with specific characteristics by the use of
whose semen they hope to give birth to a particular type of child. We do not
therefore want detailed descriptions of donors to be used as a basis for
choice, but we believe that the couple should be given sufficient relevant
information for their reassurance. This should include some basic facts about
the donor, such as his ethnic group and his genetic health.

Is it wrong for spouses to choose each other partly with a view to
their genetic contribution to children of the marriage? Is that an
analogous issue? If the donor's 'ethnic group' is a 'basic fact' then
why isn't his height, hair or eye colour, intelligence and so on? Is the
Report pandering to racial prejudice or is it just being sensible?

Committees

Although, as you will have gathered, I have grave reservations
about much of the Warnock Report, the document ends with the
beginnings of a good idea and one which is very relevant for our

purposes. It is concerned with establishing an appropriate structure for keeping important moral/legal dilemmas under review. As yet, we have no single suitable forum for structuring debate, educating a fascinated public and providing authoritative guidance on these vital issues.

The courts, for instance, can only provide sporadic review of problems, depending on the vagaries of litigation. The courts can only respond to, rather than anticipate events. Nor is the traditional English court procedure appropriate to consider the vast amount of scientific, medical, moral and economic evidence which is germane to, say, the question of allocating kidney dialysis machines. The long-running Gillick saga illustrates another disadvantage, of course: as appeal follows appeal there is often confusion and uncertainty.

The Warnock Committee could have been a more encouraging model but, apart from the faults in its reasoning above, it was only an *ad hoc* body, set up to consider a particular set of issues and now disbanded. Over two years the Committee built up some expertise in the area and received evidence from some 250 organisations and about 700 members of the public. Its Report, however flawed, has stimulated great debate and interest. A record two million people have been spurred to sign a petition in favour of a private member's bill[3] (albeit one opposing the Warnock majority view).

Indeed the number and type of groups interested in the problems facing Warnock are themselves significant. The Appendix to the Report is perhaps the most interesting part. It lists organisations and individual experts who gave evidence to the Committee. Apart from civil service briefing and the Committee's own research, this is the key to the information on the basis of which the Inquiry came to its conclusions. The Appendix is a fascinating list of concerned bodies. Some idea of the variety of viewpoints can be gleaned from an alphabetical selection of eye-catching titles: — Action for Lesbian Parents, British Toxicology Society, Campaign for Homosexual Equality – Tyneside Group, Donors' Offspring, Episcopal Church in Scotland, Free Church of Scotland, Guild of Catholic Doctors, High Court of Justice – Family Division, Institute of Marital Studies, Justices' Clerks' Society, Knights of St Columbanus – Northern Area Committee, League of Jewish Women, Mothers' Union, National Association of Ovulation Method Instructors, Presbyterian Church of Ireland, Responsible Society, Science Fiction Foundation, Trades Union Congress, United Kingdom Islamic Mission, Voluntary Council for Handicapped Children, West Indian Standing Conference and the Yorkshire Pro-Life Co-ordinating Committee. Even this wide range of organisations did not satisfy the Committee which recorded 'with regret that we did not receive evidence from as wide a range of minority and special interest groups as we would have liked, despite our best endeavours'.

Faced with the array of evidence, doubtless much of it conflicting, no-one should underestimate the task facing Warnock. Moreover, it is a salutary experience to try to formulate *any* recommendation on any of its controversial subjects when working as a group.

This, in turn, makes us think about the composition of the Committee.[4] Those who have devoted considerable professional attention, perhaps as doctors, scientists, philosophers, theologians, lawyers, social workers, to the topic in hand are unlikely to be picked as members of an official Committee because their views are known in advance and known to conflict. Yet this is only a problem if the Committee is expected to produce unanimous or majority recommendations, whereas a far more useful role for such Committees is to pave the way for public and Parliamentary debate by presenting, in a detailed but comprehensible way, the best arguments for and against the plausible alternative course of action. The individual members' preferences do not need to be recorded. It is for the rest of society to decide which way to proceed on the basis of the experts' analysis of the issues.

Now the Warnock Committee itself is not to blame for following its terms of reference which directed it to make recommendations. No doubt this helps the government of the day pass the buck of responsibility for controversial proposals, but it diverts the Inquiry's energy away from the arguments and towards a vagueness which can command wider assent. Moreover, it makes the government wary of appointing experts. Equally regrettable is the undue emphasis on what one group of sixteen would themselves prefer, instead of using their expertise to enlighten and involve all of us.

What we need now is a Super-Warnock: a permanent body to keep under review the whole range of issues in medical law and ethics. Given time, such a body would be able to produce suggested Codes of Practice covering areas such as *in-vitro* fertilisation, treatment of the young, allocation of scarce resources within the Health Service and the requirements of a sensible doctrine of informed consent.

Within its own narrow field, Warnock saw the need for such an authority. Indeed the Committee regarded the establishment of a new statutory authority with advisory and executive functions as 'by far the most urgent' of its recommendations. The *raison d'être* of its executive licensing function might well disappear if Parliament eventually bans all experiment on embryos. Nevertheless we should rescue the advisory, monitoring role and expand it to cover all the questions of medical ethics which so concern us. It may be currently unfashionable to suggest new quangos but very occasionally this is just what is required. A permanent advisory committee would fulfil a need which various forms of surrogate quanghood (such as the courts, administrative fiats and *ad hoc* committees) cannot satisfactorily meet.

Who would oppose such a body? The Government would no doubt baulk at large expenditure on a secretariat, so the new quango might have to rely initially on the existing infrastructure of research for some of its information. Nevertheless, the Government seems ready to accept the principle behind Warnock's proposed authority.

The medical establishment might be tempted to oppose the quango, but any legitimacy in that position has been undermined by their reluctant endorsement of Warnock's proposal for a licensing body. Doctors and researchers would have accepted that as the price for public acquiescence in their experiments. The public, however, may seize on that concession, without necessarily allowing experiments, in order to create a more general review body which would work in everyone's interests, helping patients and doctors alike by extensive and expert consideration of their ethical dilemmas.

What would the new quango do? Ignoring the specific references to *in-vitro* fertilisation, Warnock's explanation is a good one: 'We believe it should issue general guidance, to those working in the field, on good practice . . . and on the types of research which . . . it finds broadly ethically acceptable. It should also offer advice to Government on specific issues as they arise, and be available to Ministers to consult for specific guidance. As part of its responsibility to protect the public interest, it should publish and present to Parliament, an Annual Report'.

Who would be on the committee? Warnock again has the answer. 'The new body will need access to expert medical and scientific advice. We would therefore envisage a significant representation of scientific and medical interests among the membership. It would also need to have members experienced in the organisation and provision of services. However, this is not exclusively or even primarily, a medical or scientific body. It is concerned essentially with broader matters and with the protection of the public interest. If the public is to have confidence that this is an independent body, which is not to be unduly influenced by sectional interests, its membership must be wide-ranging and in particular the lay interests should be well represented'. Within the term 'lay', I would include experts in medical law and ethics.

Now the composition of such a body is crucial. At the very least, we can all accept that if the Government had selected sixteen research scientists working on *in-vitro* fertilisation, it would have got a recommendation to allow embryo experiments, whereas if it had chosen sixteen Catholic bishops, the recommendation would have been against experimentation on embryos.

Until recently the USA had a Super-Warnock — the inelegantly titled but otherwise admirable President's Commission for the Study of Ethical Problems in Medicine and Biomedical and Behavioural Research. Building on that model we could surely construct an institution which was able to tackle the important task of establishing codes of practice on medical ethics in a systematic and informed way. Above all, we need to have guidelines *before* doctors and researchers face the moral dilemmas directly and this in an area where, as we have seen 'both medical science and opinion within society may advance with startling rapidity'. Although much of the Warnock Report is rightly being criticised, the Committee did have the beginnings of a good idea in recommending an advisory body. That suggestion should be developed. It would be a tragedy if the embryo of a much-needed innovation is thrown out with the Warnock bath-water.

Contraception, Children's Rights, Parents' Rights: Gillick

9

Victoria Gillick took the Department of Health and Social Security to court over its guidance to doctors on the provision of contraceptive advice and treatment to young girls. The ensuing legal saga[1] was a prime example of disputes over law and morals.

The DHSS memorandum had advised doctors that they could offer under-sixteen year old children contraceptive advice and treatment without parental consent, albeit only in 'most unusual . . . exceptional cases'.

The relevant parts of the circular read as follows:

There is widespread concern about counselling and treatment for children under 16. Special care is needed not to undermine parental responsibility and family stability. The Department would therefore hope that in any case where a doctor or other professional worker is approached by a person under the age of 16 for advice in these matters, the doctor, or other professional, will always seek to persuade the child to involve the parent or guardian (or other person *in loco parentis*) at the earliest stage of consultation, and will proceed from the assumption that it would be most unusual to provide advice about contraception without parental consent.

It is, however, widely accepted that consultations between doctors and patients are confidential; and the Department recognizes the importance which doctors and patients attach to this principle. It is a principle which applies also to the other professions concerned. To abandon this principle for children under 16 might cause some not to seek professional advice at all. They could then be exposed to the immediate risks of pregnancy and of sexually-transmitted disease, as well as other long-term physical, psychological and emotional consequences which are equally a threat to stable family life. This would apply particularly to young people whose parents are, for example, unconcerned, entirely unresponsive, or grossly disturbed. Some of these young people are away from their parents and in the care of local authorities or voluntary organisations standing *in loco parentis*.

The Department realizes that in such exceptional cases the nature of any counselling must be a matter for the doctor or other professional worker concerned and that the decision whether or not to prescribe contraception must be for the clinical judgment of a doctor.

Mrs Gillick sought declarations that this circular was unlawful and that her area health authority should guarantee that her daughters would not be given contraceptive advice and treatment without her consent. She lost at first instance, won in the Court of Appeal and lost 3–2 in the House of Lords.

Overall, five judges (three in the Court of Appeal and two in the House of Lords) had ruled for her and only four (the High Court judge and three Law Lords) were against her. She lost because she failed narrowly at the ultimate stage in the House of Lords.

The judicial disagreement might suggest that this was a finely balanced controversy over law and morals. The Gillick litigation certainly illustrates the idea that disputes on law and morals turn on (1) different impressions as to factual consequences of alternative legal options and (2) different moral values. We will begin our examination of the Gillick controversy by separating these two areas of disagreement. Then we will look at the reasoning of the House of Lords, before trying to assess the significance of the dispute for our purposes.

Unlike the Warnock inquiry into how the law *ought* to respond to a problem, Mrs Gillick was apparently seeking a declaration of what the law already *was*. But, in practice, the judges may have felt that the law on her topic was unclear or unsettled, so that they had a discretion in their decision-making. Certainly, the Gillick case shows that we should not assume all judges think alike. They should not all be typecast as conservatives. Or, at least, this case shows that there is an incoherence in the liberal/conservative dichotomy often employed by commentators. Was the final decision conservative in deciding for the establishment, the government department? Or was it anti-conservative in deciding 'against' parents' rights and in favour of their children's rights to contraception?

Factual Predictions

The practical case *against* Gillick was never better expressed than by Lord Justice Parker, in the course of his leading judgment from the Court of Appeal's decision *for* Gillick:

I fully appreciate that information to the parent may lead to family trouble and that knowledge that going to the doctor involves disclosure to parents may deter others from seeking advice and treatment with, possibly, highly undesirable or even tragic results. A parent who for example, had fought long and hard for the rights which [Mrs Gillick] seeks and had won the battle might thereafter wish that she had never fought it, for it might lead to pregnancy, a back-street abortion and even death.

Another related, practical reason why Mrs Gillick's litigation may be regarded as misguided and self-defeating, is that the Court of Appeal decision might well have diminished, rather than increased, parental involvement. Before the Court of Appeal's declaration, according to this argument, doctors had generally worked in favour of parental involvement by overcoming girls' initial reluctance to talk to their parents about contraception. The vast majority of girls who had consulted a doctor in preference to their parents were eventually persuaded by the doctor to discuss the matter with their parents. But between the Court of Appeal's decision and its reversal by the House of Lords, those girls were scared of approaching a doctor at all, so the doctor had no opportunity to suggest that the girls' worries about parental involvement might be misplaced.

A few may have refrained from sexual intercourse, although it seems that most under-sixteen year olds who approach doctors for contraceptive treatment are already sexually active. A more likely response was for those girls to use contraceptives which can be bought across the counter from chemists, or to deal in a black market in contraceptives with attendant health dangers.

The other option open to these girls — of continuing to have sexual intercourse without contraception — may have led to pregnancy and to the problems of early motherhood or abortion. This taks us back to Lord Justice Parker's warning.

Hence, as I am suggesting is often the case in such law and morality controversies, the opposing views are to a large extent opposed because of different guesses as to the impact of alternative laws. In other words, different intuitions as to the practical consequences of a ruling one way or the other, are often what separate people. Would Mrs Gillick's preferred law lead to more parental involvement or less? Would it lead to more or to less early sexual activity? If we disapprove of under-age sex, abortions or young motherhood, we can still disagree as to how to pursue our disapproval most effectively. As Lord Justice Eveleigh observed, on the one hand it is said

If children think that their parents will be involved they will not come for help. Not only will they not seek contraceptive advice but they will hesitate, and thus delay, to seek advice if pregnant or after contracting a disease.

On the other side it is said that there is another way to avoid pregnancy, namely by abstinence; and this is the only 100% guarantee against pregnancy and disease. The availability of secret medical advice undermines the efforts of the parent to bringing the child up with proper moral standards and encourages promiscuity.

So much for the practicalities. Now what about the *morality* of providing fifteen-year old girls with contraceptive advice or treatment in the absence of parental consent?

Moral Values

In morals, age cannot conclusively determine such a matter. The key issue is, instead, the ability of a person to understand the issues involved in contraceptive treatment. Only in this way can medical ethics respect the autonomy of the individual patients. If someone, whether aged fifteen or fifty, is incapable of understanding the nature and consequences of a course of treatment, *then* we can consider if a third party can supply a valid consent.

It should be noted, however, that even in such circumstances, good medical ethics would require the consent to be given or withheld in the interest of the patient, not in the interest of the proxy. In ethics, therefore, our central concern should be the maturity, not the calendar age, of a patient. This respect for autonomy is the consequence of recognizing the equal moral worth of us all. It is the bedrock of good medical ethics.

Mrs Gillick lost the first hearing of her case before Mr Justice Woolf in the High Court. Mr Justice Woolf accepted the argument which we have just put forward:

The fact that a child is under the age of 16 does not mean automatically that she cannot give consent to any treatment. Whether or not a child is capable of giving the necessary consent will depend on the child's maturity and understanding, and the nature of the consent which is to be required. The child must be capable of making a reasonable assessment of the advantages and disadvantages of the treatment proposed, so the consent if given can be properly and fairly described as a true consent.

Another sensible snippet from that judgment is Mr Justice Woolf's rejection of the language of 'parental rights', since 'the interest of parents, I consider, are more accurately described as responsibilities or duties'. This reflects a paradox running through the Gillick campaign. In the good old days, the Moral Right accused the Immoral Left of emphasising rights and neglecting duties, thus ushering in the permissive society. Now it is Gillick and friends who are insisting on parental rights and thus misunderstanding the parental role.

Can the Law Rely on a Flexible Approach?

In the Court of Appeal, however, Mr Justice Woolf's approach was rejected. Lord Justice Parker seems to have driven a wedge between ethics and the law by saying, 'A child may be of sufficient understanding and intelligence to give a consent before, or not until after, it has attained whatever may be the fixed age, but if there be no such age then neither parent, child nor strangers will know what their respective positions are'.

In fact, the law is quite used to evaluating the viability of consent or understanding in individual cases, without relying on arbitrary age-limits. Courts have to decide whether ten to thirteen-year olds knew that they were doing wrong, before holding them responsible for crimes. Again, the courts are well used to judging maturity when determining whether a child has a sufficient capacity for understanding truth to give evidence under oath.

Most importantly, however, the House of Lords had recently considered the question of capacity to consent in the context of the criminal issue as to whether a father can kidnap his son. Lord Brandon had spoken for the Law Lords in declaring: 'It must, I think, be a question of fact for the jury whether the child concerned has sufficient understanding and intelligence to give its consent'.

And another germane statement of principle came from Lord Denning in another case where he was talking about the meaning of describing a fifteen-year-old as 'in the custody of parents'. He said that the legal right of parents ends when the child becomes eighteen, but before then, 'it is a dwindling right which the courts will hesitate to enforce against the wishes of the child, and the more so the older he is. It starts with a right of control and ends with little more than advice'.

The Law Lords

It is important to unravel the different approaches adopted by the five Law Lords. Three were in favour of allowing the DHSS appeal, but two dissented.

The Majority

Lords Fraser, Scarman and Bridge seemed to have a common approach. They sought out the principle underlying previous case law on the general relationship between children and parents and then applied it to the specific context of whose consent is required for contraceptive advice or treatment. In Lord Scarman's words, the principle is that 'parental rights are derived from parental duty' and the 'dwindling right' of a parent as the child grows older 'yields to the child's right to make his own decision when he reaches a sufficient understanding and intelligence to be capable of making up his own mind on the matter requiring decision'.

This stage will vary from child to child and from decision to decision. The law can and should acknowledge the reality of different rates of development towards maturity.

So Lord Scarman's method is to focus on the young person herself. If she can consent to contraceptive advice, the law will respect her autonomy. If she cannot, then the law will look for a proxy, who will usually be a parent. But parental consent is not the first option and whenever it is necessary it should be exercised in the child's best interest.

Having established the general principle, contraceptive treatment then needs to be located within this wider context. Lord Fraser sets out five guidelines which doctors must follow before they can legally offer a girl contraceptive advice or treatment without parental consent.

The doctor must be satisfied

(1) that the girl (although under 16 years of age) will understand his advice; (2) that he cannot persuade her to inform her parents or to allow him to inform the parents that she is seeking contraceptive advice; (3) that she is very likely to begin or to continue having sexual intercourse with or without contraceptive treatment; (4) that unless she receives contraceptive advice or treatment her phsyical or mental health or both are likely to suffer; (5) that her best interests require him to give her contraceptive advice, treatment or both without the parental consent'.

Lord Fraser follows his five rules immediately with the reassurance that they 'ought not to be regarded as a licence for doctors to disregard the wishes of parents on this matter whenever they find it convenient to do so'.

But to understand what his first point (about understanding advice) entails, and why it is a difficult hurdle for the girl and the doctor to surmount, we must turn to Lord Scarman's judgment. He emphasized that

there is much that has to be understood by a girl under the age of 16 if she is to have legal capacity to consent to such treatment. It is not enough that she should understand the nature of the advice which is being given: she must also have a sufficient maturity to understand what is involved.

There are moral and family questions, especially her relationship with her parents; long-term problems associated with the emotional impact of pregnancy and its termination; and there are risks to health of sexual intercourse at her age, risks which contraception may diminish but cannot eliminate. It follows that a doctor will have to satisfy himself that she is able to appraise these factors before he can safely proceed upon the basis that she has at law capacity to consent to contraceptive treatment.

That is the law, but why is it the law? The judges explained that this was in keeping with the long-standing tradition of the law. It was the Court of Appeal which was out of step with legal principle. Thus Lord Scarman says,

the underlying principle of the law was exposed by Blackstone and can be seen to have been acknowledged in the case law. It is that parental rights yields to the child's right to make his own decisions when he reaches a sufficient understanding and intelligence to be capable of making up his own mind on the matter requiring decision.

Moreover, the law accorded with common sense. Lord Fraser explains how this legal principle reflected social reality:

It is, in my view, contrary to the ordinary experience of mankind, at least in western Europe in the present century, to say that a child or a young person remains in fact under the complete control of his parents until he attains the definite age of majority, now 18 in the United Kingdom, and that on attaining that age he suddenly acquires independence. In practice most wise parents relax their control gradually as the child develops and encourage him or her to become increasingly independent. Moreover, the degree of parental control actually exercised over a particular child does in practice vary considerably according to his understanding and intelligence and it would, in my opinion, be unrealistic for the courts not to recognise these facts.

The Dissents

Lords Brandon and Templeman did not join in the majority's quest for legal principle. This was particularly curious in the case of Lord Brandon for it was his judgment in an earlier case which had convinced the majority that a young person's capacity to consent depended on the individual's own understanding and intelligence rather than on calendar age. It seems that the dissentients shared Mrs Gillick's worries about the practical consequences of the majority view and they therefore ignored the search for legal principle.

Lord Brandon went straight to the narrower question of whether the doctor would be guilty of a criminal act if he prescribed contraceptives to under-sixteens and gave an affirmative answer.

For Lord Brandon, the question whether contraceptive advice or treatment is lawful 'appears to me to be one of public policy, the answer to which is to be gathered from an examination of the statutory provisions which Parliament has enacted from time to time in relation to men having sexual intercourse with girls either under the age of 13 or between the ages of 13 and 16'.

Lord Brandon does not explain why he ignores the majority's search for legal principle. Nor does he explain why he disagrees with the majority when they rely on Lord Brandon's own landmark decision in the kidnapping case. The dissenting Lord Brandon is indeed, as I have said, in the curious position of being the principal authority for the majority.

In any event, the statutes which Lord Brandon relies on are those which make sexual intercourse with an under-sixteen year old girl a serious criminal offence for the *man*. Now what is the public policy behind this? As Lord Brandon admits, it is 'that the relevant statutory provisions have been enacted by Parliament for the purpose of protecting the girl from herself'.

But how does Lord Brandon explain another statutory provision: s5(1) of the National Health Service Act 1977? This states that,

It is the Secretary of State's duty . . . to arrange, to such extent as he considers necessary to meet all reasonable requirements in England and Wales, for the giving of advice on contraception, the medical examination of persons seeking advice on contraception, the treatment of such persons and the supply of contraceptive substances and appliances.

This has no mention of age-limits. If we are really interested in Parliament's public policy, then this is surely game, set and match to the DHSS. But, no. Lord Brandon says that he 'would interpret the expression 'persons' in s5(1)(b) as not including girls under 16'!

Having rewritten the governing statute so as to deem under-sixteen year old girls to be non-persons, Lord Brandon notes the impact of his approach:

My Lords, great play was made in the argument before you of the disastrous consequences for a girl under 16 of becoming pregnant as a result of her willingly having unlawful sexual intercourse with a man. I am fully conscious of these considerations, but I do not consider that, if the views which I have so far expressed are right in law, those considerations can alter the position.

Lord Brandon's strange use of public policy thus brings him to a view which was more extreme than Mrs Gillick's. 'This is because', as his Lordship explains, 'on the view which I take of the law, making contraception available to girls under 16 is unlawful, whether their parents know of and consent to it or not'.

Two of the majority Law Lords devote a paragraph each to rebutting Lord Brandon's arguments. Lord Scarman's magisterial opinion dismisses arguments based on alleged criminality or public policy as 'surprising'. In his Lordship's view,

It cannot be said that there is anything necessarily contrary to public policy in medical contraceptive treatment if it be medically indicated as in the interest of the patient's health; for the provision of such treatment is recognised as legitimate by Parliament: see s5 of the National Health Service Act 1977.

Of course it *can* be said, as Lord Brandon has shown, but otherwise Lord Scarman has summarised the position succinctly.

Lord Bridge concludes,

On the issue of public policy, it seems to me that the policy consideration underlying the criminal sanction imposed by statute on men who have intercourse with girls under 16 is the protection of young girls from the untoward consequences of intercourse. Foremost among these must surely be the risk of pregnancy leading either to abortion or the birth of a child to an immature and irresponsible mother. In circumstances where it is apparent that the criminal sanction will not, or is unlikely to, afford the necessary protection it cannot, in my opinion, be contrary to public policy to prescribe contraception as the only effective means of avoiding a wholly undesirable pregnancy.

I suspect that the reason why Lord Brandon took such a strained view of the policy implications of the statutory material was that he had, perhaps subconsciously, convinced himself that the consequences argument pointed in Mrs Gillick's favour.

Lord Templeman eschewed case-law and produced an opinion which reads like a speech by Mrs Gillick. The most memorable line is that: 'There are many things which a girl under 16 needs to practise but sex is not one of them'. He speculated about the effect of allowing contraceptive treatment without parental consent. No factual evidence on this point was available to any of the courts hearing the case.

Lord Templeman's dissenting judgment seems to take the view that no girl under the age of sixteen *could* understand all the factors involved in the request of contraceptive advice and treatment. But note that such a girl would be able to consent to some other forms of treatment on his analysis.

I accept also that a doctor may lawfully carry out some forms of treatment with the consent of an infant patient and against the opposition of a parent based on religious or any other grounds. The effect of the consent of the infant depends on the nature of the treatment and the age and understanding of the infant. For example, a doctor with the consent of an intelligent boy or girl of 15 could in my opinion safely remove tonsils or a toublesome appendix. But any decision on the part of a girl to practice sex and contraception requires not only knowledge of the facts of life and of the dangers of pregnancy and disease but also an understanding of the emotional and other consequences to her family, her male partner and to herself. I doubt whether a girl under the age of 16 is capable of a balanced judgment to embark on frequent, regular or casual sexual intercourse fortified by the illusion that medical science can protect her in mind and body and ignoring the danger of leaping from childhood to adulthood without the difficult formative transitional experiences of adolescence. There are many things which a girl under 16 needs to practise but sex is not one of them.

Application of the Law

We are presented, therefore, with an interesting problem for psychologists and others to research. Who is right — the majority Law Lords or Lord Templeman — in their analysis of the competence of a young girl to understand these issues? Lord Scarman observed that this is a factual dispute: 'It will be a question of fact whether a child seeking advice has sufficient understanding of what is involved to give a consent valid in law'.

There is a further practical problem for doctors to resolve. Even if it is possible in principle for a girl to consent, how can a doctor test her capacity in the limited time available? Again, there is a responsibility on doctors to devise means of giving information and asking question so as to determine a girl's capacity to consent. I do not pretend that this is an easy task although doctors ought to be testing capacity to consent in a variety of other circumstances, such as in cases of senile dementia or psychiatric cases.

Suppose a particular girl does not have the capacity to consent. What can the doctor do? The normal proxy will be a parent. How should the parent exercise his or her power of consent? The vital principle to bear in mind here is that a proxy has the duty to decide *in the best interests of the child* and not in the best interests of the proxy where those might differ. One of the reasons why the parent acts as proxy is because, in Lord Fraser's words, 'in the overwhelming majority of cases the best judges of a child's welfare are his or her parents'. But this will not be true in all cases. Even the dissenting

Lord Templeman could think of cases where it would not be necessary for the doctor to get parental consent in advance of treatment:

The doctor is entitled in exceptional circumstances and in emergencies to make provision, normally temporary provision, for contraception but in most cases would be bound to inform the parent of the treatment. The court would not hold the doctor liable for providing contraceptive facilities if the doctor had reasonable grounds for believing that the parent had abandoned or abused parental rights or that there was no parent immediately available for consultation or that there was no parent who was responsible for the girl.

Lord Scarman makes clear the path for doctors to follow: first, look at the child's capacity to consent; second, look at the parent's suitability to act as proxy; third, in exceptional circumstances where the parent cannot or should not be consulted, then the doctor must act as proxy. His Lordship puts it in this way:

Until the child achieves the capacity to consent, the parental right to make the decision continues save only in exceptional circumstances. Emergency, parental neglect, abandonment of the child, or inability to find the parent are examples of exceptional situations justifying the doctor proceeding to treat the child without parental knowledge and consent: but there will arise, no doubt, other exceptional situations in which it will be reasonable for the doctor to proceed without the parent's consent.

When the doctor is making the decision, it is again perfectly clear that he must act in the best interests of his patient. Lord Fraser's guidelines are quite explicit on this. His Lordship stresses that 'the solution depends upon a judgment of what is best for the welfare of the particular child'. The 'best-interests' criterion is also highlighted in the passage which immediately precedes the five guidelines. Lord Fraser is convinced of the

desirability of the doctor being entitled in some cases, in the girl's best interest, to give her contraceptive advice and treatment if necessary without the consent or even the knowledge of her parents. The only practicable course is, in my opinion, to entrust the doctor with a discretion to act in accordance with his view of what is best in the interests of the girl who is his patient.

Now what are in the best interests of the girl? Here there might be a disagreement as to whether it is best to withhold contraceptive treatment in the hope that this will discourage premature sexual activity, or to provide contraception as a means of preventing some consequences of such activity. But even Lord Templeman seems to have admitted, in the passage quoted above, that it could be in the best interests of a girl for the doctor to prescribe contraceptives in some circumstances, and the majority clearly envisages such action, subject to Lord Fraser's guidelines.

So attention now turns to the medical profession's response to Gillick. Lord Templeman seems to take a pessimistic view while Lords Fraser and Scarman have nothing but praise for, and confidence in, doctors. Lord Templeman conjures up the picture provided by Mrs Gillick of an 'inexperienced doctor in a family planning clinic, exuding sympathy and veiled in ignorance of the girl's personality and history, [providing] contraceptives as if they were sweets withheld from a deprived child by an unfeeling parent'.

Lord Scarman, on the other hand, is optimistic:

I accept that great responsibilities will lie on the medical profession. It is, however, a learned and highly trained profession regulated by statute and governed by a strict ethical code which is vigorously enforced. Abuse of the power to prescribe contraceptive treatment for girls under the age of 16 would render a doctor liable to severe professional penalty. The truth may well be that the rights of parents and children in this sensitive area are better protected by the professional standards of the medical profession than by 'a priori' legal lines of division between capacity and lack of capacity to consent since any such general dividing line is sure to produce in some cases injustice, hardship and injury to health.

The Law Lords have given doctors the opportunity and responsibility to demonstrate that it is Lord Scarman and not Lord Templeman who has grasped the truth.

The Significance of Gillick for Law and Morals

The Gillick saga illustrates that controversies over law and morality often turn on different hunches as to facts and conflicting estimates of consequences. Mrs Gillick, together with Lords Brandon and Templeman, considers that premature sexual activity is best discouraged, and parental involvement is best encouraged, by the law refusing to accept a child's consent as sufficient for the provision of contraceptives.

Others feel that premature sexual activity has to be discouraged by different means, chiefly educational. They believe that the absence of contraceptives will not act as a deterrent. On the contrary, it will merely cause additional problems such as abortion or early motherhood. Moreover, Mrs Gillick's opponents say that a victory for her would have *decreased* parental involvement in the real world. In the vast majority of cases which arose before the litigation, they claim, doctors were able to persuade initially reluctant girls to confide in their parents. Since the Court of Appeal decision, the girls concerned had not given doctors that opportunity.

Although Lords Brandon and Templeman were prepared to rely on their hunches, the courts are not an appropriate forum in which to evaluate these competing claims, at least under present circumstances. This is why the majority of Law Lords emphasized legal principle and thought that, in Lord Bridge's words, the court should avoid 'expressing *ex cathedra* opinions in areas of social and ethical controversy in which it has no claim to speak with authority'.

A standing commission on medical law and ethics, a Super-Warnock, is long overdue. In the absence of such a body, however, the courts deserve praise for filling the vacuum. Lord Scarman thought that Mrs Gillick too deserved praise for raising the issue.

Of course many are glad that Mrs Gillick was defeated. But it is not fair to criticize her for trying to enforce her morality through the law. Both sides of this dispute and all sides of all disputes on law and morals are trying to enforce their morality through the law. This applies to liberals who want the law to reflect liberalism just as much as it applies to Mrs Gillick who wants the law to reflect 'Gillickism'. The argument should be about the respective merits of the rival creeds, not based on the spurious belief that only one side is trying to impose its vision of morality.

Those of us who find the House of Lords' decision wholly persuasive are not, whatever Mrs Gillick may think, thereby condoning premature sexual activity. On the contrary, I think it is our duty to convince the young people at risk of the moral, psychological, emotional, physical and social harm involved. But the way forward is through education rather than an attempt to change the law on young people's access to contraception.

For our purposes, one of the most interesting aspects of Mrs Gillick's campaign is that she was regularly described as a 'Roman Catholic mother of ten' and that her campaign was stereotyped as a crusade by the Roman Catholic wing of the Moral Majority. Yet she was critical of other Catholics, including the bishops, whenever she sensed a lack of whole-hearted support. She seemed to assume that the bishops should automatically endorse her every legal move. While the post-Vatican II Roman Catholic church has seen a shift by the laity away from unconditionally following the bishops in every way, it is far from having become a Church in which the bishops are expected unconditionally to follow one section of the laity.

Mrs Gillick did seem to have difficulty in separating law and

morals. Feeling that the DHSS memorandum was morally unacceptable, she tried to have it declared legally unacceptable. In this, she
ultimately failed. Some people disagreed with her litigation on the
ground that, if successful, it would have been counter-productive.
Others disagreed because they had a different scheme of moral
values, placing greater emphasis on the autonomy of young people
rather than on the rights of parents over their children.

Other Roman Catholics could be numbered among those dis-
agreeing with her litigation. They might take a different view of the
practical consequences or they might be suspicious, with one eye on
the abortion debate, of allowing parental views on their children's
future to be given absolute moral or legal priority. Once again, then,
there is no reason to assume that all Roman Catholics will think alike
on law and morals.

The Gillick case, finally, illustrates that the law has to respond to
developments in society; it shows how the courts should set about
that response; and it demonstrates the limitations of law as an
answer to moral problems.

Law and morals evolve. The law needs a catalyst to alert it to any
changes. Ironically, Mrs Gillick was the catalyst here for a striking
affirmation by the majority Law Lords of *children*'s rights. To this
extent, as Lord Scarman observed, Mrs Gillick, 'even though she
may lose the appeal, has performed a notable public service in
directing judicial attention to . . . problems . . . of immense consequence to our society'.

Lord Scarman concludes our survey of the Gillick controversy by
explaining why the problem arose and how the law can be developed accordingly:

Three features have emerged in today's society which were not known to our
predecessors: (1) contraception as a subject for medical advice and treatment;
(2) the increasing independence of young people; and (3) the changed status
of women. In times past contraception was rarely a matter for the doctor; but
with the development of the contraceptive pill for women it has become part
and parcel of everyday medical practice . . . Furthermore, women have
obtained by the availability of the pill a choice of life-style with a degree of
independence and of opportunity undreamed of until this generation and
greater, I would add, than any law of equal opportunity could by itself effect.
'The law ignores these developments at its peril. The House's task, therefore,
as the supreme court in a legal system largely based on rules of law evolved
over the years by the judicial process is to search the overfull and cluttered
shelves of the law reports for a principle or set of principles recognised by the

judges over the years but stripped of the detail which, however appropriate in their day, would, if applied today, lay the judges open to a justified criticism for failing to keep the law abreast of the society in which they live and work.

Through the good sense of the majority Law Lords, albeit without the benefit of evidence, the law ultimately responded.

The law is not going to use the threat of pregnancy, followed by early motherhood or abortion, as a means of scaring young girls into sexual abstinence, although the criminal law will do its best to punish those men who have sexual intercourse with girls aged less than sixteen. But if we think that premature sexual activity is undesirable for a variety of moral, physical, emotional and psychological reasons, then it is our responsibility to put that view across to the young people at risk. Education and morals may be a more useful conjunction here than law and morals.

10

Medical Paternalism:
Informed Consent

Consent was at the centre of another 1985 House of Lords' decision on medical law.[1] Mrs Sidaway underwent an operation to relieve pain in her shoulders and neck. There was a very small risk (less than one per cent) of a very severe injury if the spinal cord was damaged by the surgery. The risk materialised and Mrs Sidaway was left severely disabled. The operation was not carried out negligently, so Mrs Sidaway could not bring a legal action on that ground.

But Mrs Sidaway claimed that she had not been informed of this particular risk and that, if she had been so informed, she would not have agreed to the operation. As the surgeon had died between the operation and the legal action, it was difficult to establish what he had said to Mrs Sidaway. Nevertheless, the judge was 'satisifed that he did not refer to the danger of cord damage'. Mrs Sidaway sought to invoke the doctrine of 'informed consent' which is more popular in American legal circles. She argued that the surgeon ought to have warned her of all the risks associated with the operation. Without that knowledge, she was unable to give an 'informed consent' to the operation.

Consent is an important notion for lawyers. The general idea here is that nobody has a right to touch my body without my agreement. The principle is a classic application of the harm-to-others formula. Physical interference counts as 'harm' unless I consent to it. We often do consent to others touching our bodies for good reason — because we want a haircut or we want to play rugby or we want the doctor to examine us. If somebody rugby-tackles me in the street, however, I will probably be able to sue him for the wrong (or, in legal language, the 'tort') which he has done to me.

But surely I must know what is going on before I can validly consent to something. What exactly do I need to know? Mrs Sidaway's argument was that to give a genuine consent she would have had to know the *risks* involved.

A truly 'informed' consent would also require an understanding of the *alternatives* to a proposed course of action.

Clearly, the idea of 'informed consent' is based on the notion of autonomy which we have already encountered. Autonomy rests on rationality. It is difficult to act rationally in the absence of relevant information. It is on the basis of the benefits, risks and alternatives that we formulate reasons for a course of action.

Mrs Sidaway, then, is relying on the value of autonomy. What possible explanation, other than inadvertence, could there be for the surgeon failing to disclose fully the risks of an operation? Quite simply, the surgeon believes that the patient would, on balance, benefit from the operation (even allowing for the risks) but that disclosure of the risks might worry the patient into withholding her consent. This is paternalism.

In the Gillick controversy, as we have seen, many doctors (and ultimately the law) preferred the values of young people's autonomy and medical paternalism to parental paternalism. We are accustomed to medical paternalism. Doctor knows best. But is this right? The Sidaway case sets the scene for a discussion of paternalism as it relates to informed consent.

All nine judges who heard Mrs Sidaway's actions dismissed her claim for damages based on her surgeon's failure to inform her of a risk in an operation. Even on the test most favourable to her, Mrs Sidaway had failed to prove her case. But the judges' debate over the proper test reflects a difference in attitudes towards law and morality. The simple choice of autonomy or paternalism was presented starkly. Facts and precedents could not obscure the issue in Sidaway because the facts were so sketchy as to provoke Lord Diplock to observe that: 'it is the very paucity of facts in the evidence that makes it possible in my view to treat this appeal as raising a unique question of legal principle' and the previous case law contained no English appellate court decision.

Picking up that last point Lord Scarman proceeded to set out the issue as follows: 'the case is plainly of great importance. It raises a question which has never before been considered by your Lordships' House: has the patient a legal right to know, and is the doctor under a legal duty to disclose, the risk inherent in the treatment which the doctor recommends'.

Four of the Law Lords seemed to give the benefit of the doubt to the doctors. In other words, they erred on the side of paternalism. As Lord Templeman put it: 'at the end of the day the doctor, bearing in mind the best interests of the patient and bearing in mind the patient's right to information which will enable the patient to make a balanced judgement, must decide what information should be given to the patient and in what terms that information should be couched'.

Lord Scarman, on the other hand, plumped for the patient and autonomy:

in a medical negligence case where the issue is as to the advice and information given to the patient as to the treatment proposed, the available options and the risks, the court is concerned primarily with a patient's right. The doctor's duty arises from his patient's rights. If one considers the scope of the doctor's duty by beginning with the right of the patient to make his own decision as to whether he will or will not undergo the treatment proposed, the right to be informed of significant risk and the doctor's corresponding duty are easy to understand, for the proper implementation of the right requires that the doctor be under a duty to inform his patient of the material risks inherent in the treatment.

But the Sidaway decision was not simply Informed Consent 1: Uninformed Consent 4. Even the four whose inclinations led them to support doctors in the last analysis, nevertheless accepted the need for informed consent in some circumstances. Lord Diplock, for example, gave judges like himself the solace of informed consent:

When it comes to warning about risks, the kind of training and experience that a judge will have undergone at the bar makes it natural for him to say (correctly): it is my right to decide whether any particular thing is done to my body, and I want to be fully informed of any risks there may be involved of which I am not already aware from my own general knowledge as a highly educated man of experience, so that I may form my own judgment whether to refuse the advised treatment or not.

But why should only 'highly educated men of experience' have control over what is done to their bodies? Non-highly educated men and women of experience and innocence will surely, sooner or later, come to benefit from such control. Indeed Lord Bridge, with whom Lord Keith agreed, had already broadened Lord Diplock's approach to include non-judges: 'When questioned subsequently by a patient on parental consent about risks involved in a particular treatment proposed, the doctor's duty must, in my opinion, be to answer both truthfully and as fully as the question requires'. Lord Templeman agreed: 'Mrs Sidaway could have asked questions. If she had done so, she could and should have been informed that there was a risk'.

So if the Sidaway case poses a simple conflict between autonomy and paternalism, the resolution of that conflict is not so simple. The majority of the Law Lords seem to think that autonomy is not an absolute value, that individuals are not necessarily always the best judges of their own interests and that therefore paternalism by the doctor is sometimes justifiable. But when pressed, they too seem to give priority to autonomy, at least for those who have the initiative to ask about the risks of proposed treatment.

All their Lordships agreed on the principle that those who ask can expect the information on which to give a truly informed consent. And if one only has to ask what are the risks, then surely practice will gradually evolve into that question being habitually asked by patients. Bolder patients may lead the way, but it is surely not beyond the capacity of all patients to follow that path. Even if it is too much for the overawed patient to trigger the doctrine with a question, an education-based distinction between patients' rights is unsatisfactory and the law will change to devalue the importance of actually asking the magic question. Sooner or later it will be taken as read.

So Sidaway raises the question whether the law and doctors should adopt one moral position or another. The law could sanction doctors' natural tendency to be paternalistic or it could say that the autonomy of the individual patient is paramount. As we have seen, the Law Lords end up saying both. The confusion in their judgments is not surprising. This was unchartered legal territory. Moreover, the moral values at stake are not easily resolved. The context of Sidaway is important in leading us away from the preoccupation with sex and the criminal law and leading us towards a central problem in medical law and ethics. The doctor/patient relationship revolves around the requirement of consent. But Gillick on the giving of consent by proxies and Sidaway on identification of risks do not exhaust the moral and legal debate about medical consent.

Informed consent involves more than the disclosure of risks. The consent has, in addition, to be voluntary. Most importantly, informed consent also depends on the discussion and consideration of alternative courses of action. Thus a patient cannot be given the option of informed consent even if she is told all about the risks of the contraceptive pill unless she is also told about the advantages and disadvantages of other contraceptive techniques. Or, to give another example which has caused concern in the recent past, a woman suffering from breast cancer cannot give a genuine consent to the removal of her breast if she is not told about the alternative treatments which might render this drastic action unnecessary.

In the medical context, the *law*'s role in establishing a relationship based on informed consent is not necessarily the most important. Medical practice and medical education may have more say in what happens behind closed surgery doors than the law can ever hope to achieve. The law of tort only gets involved tangentially when something goes wrong and a patient seeks financial compensation. In these circumstances it is inevitably difficult for the patient to prove her case. There are problems in providing independent evidence and in finding medical experts who are prepared to criticise a colleague.

We must look elsewhere beyond the law if we want medical practice to incorporate our preferred moral approach. Whether we want patient autonomy or medical paternalism to prevail, the answer will lie in the standards of medical care laid down by the British Medical Assocation and the General Medical Council and inculcated in medical training. These bodies are in the business of making rules to guide doctors. In formulating those rules they inevitably decide questions of morality. Their decisions, in practice, may well be more important than the court's decisions, so it is important to broaden the debate about the enforcement of morality beyond law to include the whole gamut of ways in which society seeks to establish its moral precepts.

11

Resource Allocation: Kidney Dialysis

Mr Sage achieved some fame in 1985 when the Oxfordshire Area Health Authority refused to continue his kidney dialysis treatment on the grounds that he did not have 'a sufficiently high quality of life'. Although Mr Sage was subsequently given dialysis treatment in a London hospital, paid for privately by a charity, he died shortly thereafter.

Now this sad case raises another dilemma for medical ethics: the allocation of scarce resources. Or, at least, it provides a convenient peg on which to hang this long-standing problem. I add that rider because it is far from clear that Mr Sage's treatment was terminated in order to give a more deserving case the benefit of kidney dialysis. It seems that there were enough machines to go round, at least in the Oxfordshire area, but the Health Authority had doubts about Mr Sage's lifestyle and reaction to medical care.[1] His continuing treatment seemed to be disruptive to other patients and to the medical team, while its countervailing benefits were less easy to measure. So Mr Sage's case may be an unusually sad one.

Nevertheless, if we focus on the question of who should get kidney dialysis treatment where there is competition for scarce resources, then the quality of life of different patients might be one criterion for selection. But there are plenty of other ways in which we could choose to allocate kidney dialysis. Efficiency, for example, could be our standard. In which case, we might reject those who, because of their advanced age were prone to other health problems so that society might not get many hours of life from however many hours of treatment were provided. Efficiency might also lead to the rejection of those who were unlikely to keep to the rigid diet necessary for dialysis to work. This might exclude, for example, the young or even the less well-educated.

It is obvious that what could be a rational criterion, efficiency, here begins to seem immoral or unjust if it starts to exclude patients on grounds of age or especially of intelligence.

So equality could be our alternative standard. Like cases should be treated alike. But what counts as a like case? The concept of equality may well be questionable, or at least it will rest on other decisions as to the relevant criteria.

As in Sage's case the quality of life could be a different guideline. But what do we mean by the quality of life and why is it significant? One could argue that if society is paying for kidney dialysis treatment and there is only one machine between two patients then society is entitled to prefer the brilliant musician or sportsman to the 'down and out'. On the other hand one could find this repugnant. A variation on the quality of life theme would be the standard of a patient's value to society. Again, if society is paying, why not give the machine to the brain surgeon rather than to the unemployed youth?

If we find these competing standards sufficiently convincing or conflicting or awful to choose between them, we can always make the decision by abdicating the decision in giving the selection over to a lottery. But yet again raffling kidney dialysis treatment may offend our moral sensibilities.

If none of the above solutions recommend themselves we could always try to avoid the decision by insisting that society provides enough kidney dialysis machines to go round.

But the money for those extra machines must come from somewhere else in the National Health Service budget or from somewhere else in another government department's budget or at least from somewhere else. We cannot ultimately dodge the question of resource allocation. Sooner or later, we have to confront it, not only in kidney dialysis but in many other areas of medicine. This issue has recently been raised in relation to cervical smear testing, for example. Has the NHS devoted enough resources to the sufficiently regular and prompt testing of women, and has it devoted adequate resources to informing those women of the results of their tests? Or again, the much-publicised heart- and lung-transplants are sometimes criticised by those who would like to see the money spent at the less exotic end of the NHS's duties.

All of this, I hope, will seem like a trend away from law and morality towards politics and justice. So it is. Or rather law and morality and politics and justice are all part of a continuum. Questions of resource allocation are often thought of as being about political choices resting on competing notions of justice, but they are essentially about the way in which society chooses between conflicting values in its distribution of available goods. The law and morality debate should be placed within that larger question, for law is one technique by which society makes such choices, and what we choose to think of as morality constitutes one part of the whole range of our values or reasons for action.

I have argued that it is therefore important to consider the structures through which society makes such decisions. There are no simple solutions. There is no simple sentence (with respect to Mill). There is no one simple value whether it be autonomy, paternalism, equality or liberty which provides all the answers. Nor is it possible for any one individual to predict accurately the consequences of society taking a different available option. But if we do have a society in which we can all contribute to the debate about Warnock, Gillick, Sidaway and Sage by being informed and educated with the help of experts, then our society will benefit from such reappraisal of fundamental values and constant refining of predictions about the practical consequences of society's decisions. The medical context seems to be the one which has most caught the public imagination in recent times. The administrative decision in Sage and the judicial decisions in Gillick and Sidaway must inevitably stem from a limited perspective. The continuing debate over Warnock seems to me a better model, if adapted in the way I have indicated, for a resolution of such difficult matters, but it is not only in the medical field that such problems arise. Before trying to draw together some threads from this discussion, therefore, I wish now to reflect on two different contexts which seem to me to be particularly important: discrimination and freedom of expression.

12

Equality: Discrimination

Equality is a value which features regularly in discussions of law and morals. We have just seen how there may be different conceptions of what equality means as applied to a problem like the allocation of kidney dialysis machines. And the Hart–Devlin debate, of course, drew our attention to the question whether homosexuals and heterosexuals should be treated equally. This is often put in another way: is it right to discriminate against homosexuals? By reflecting on various forms of discrimination, we may be able to edge towards a clearer understanding of what we mean by equality, why we value it, and when we wish the law to enforce it. We will begin with discrimination against homosexuals, not only because of the Hart-Devlin legacy but also because the deadly disease AIDS (Acquired Immune Deficiency Syndrome) seems to have some propensity to afflict practising homosexuals.

Is discrimination against AIDS victims justifiable? Is discrimination against homosexuals, on the ground that they are more likely to become AIDS victims, justifiable? Let me set the scene with extracts from an article in *The Times* describing New York's worries about AIDS (5 September 1985):

More than 4,300 people in this city have AIDS (Acquired Immune Deficiency Syndrome), a third of all the cases in the United States, and New Yorkers are growing increasingly frightened as the fatal disease spreads . . . parents have won the support of New York's mayor in opposing the admission to school of children with AIDS. Fear of the disease has been a sigificant factor . . . Even since AIDS was first identified in the US in June 1981, many of its victims have found themselves regarded with loathing. They have become the untouchables of the 1980s, thrown out of their homes and jobs and shunned by acquaintances . . .

This fear of AIDS, its victims and its potential victims has not been allayed by assurances that, as an official New York city pamphlet puts it, 'AIDS is not highly contagious and is not spread through casual or nonsexual contact. The virus is not spread through the air or in food . . . associating with people with AIDS, or with members of high-risk groups does not pose any risk of contracting the disease'.

Do AIDS victims 'harm others'? If we believe the pamphlet, they do not pass on the disease just through attending school, yet, because others irrationally fear this might happen, they are subjected to discrimination. In a way, of course, the fear would be regarded as a 'harm', although even on such a wide definition it would seem harsh to brand the AIDS victims as *causing* such 'harm'. But if we accept that the harm-to-others test is far from being determinative, how should New York react to the AIDS problem? This may be a classic illustration of discrimination stemming from fear which in turn stems from ignorance. Dispelling the fear, rather than caving in to it would seem to be the ideal response. Trevor Fishlock's 'Letter from New York' concludes with another city's legal solution: 'Los Angeles has passed a law to protect AIDS victims from discrimination in housing, work and education. Homosexuals are concerned that the existence of the disease is increasing prejudice against them'. And well they might be.

Moreover, we need to know the facts about the origins and transmission of AIDS before we can pass authoritative judgement. In the real world, of course, legislators and administrators have to operate on the basis of partial knowledge. But the dispute about how to regard AIDS victims in the light of our present incomplete knowledge is largely about (1) what are the predicted consequences? and (2) do we place a premium on AIDS victims' liberty and rights or on the safety of others?

The second question is not *so* vital if we predict that there is no chance of normal social contact transmitting the disease. But it may still be necessary to prevent AIDS carriers from donating blood through pleas for self-restraint, administrative regulation or even legislation.

There seems little reason to object to this. There is no profound reason to let people donate contaminated blood, so there is little to weigh in the moral value balance against the clear danger.

But if, as seems likely, AIDS victims or carriers can transmit the disease not only through donating blood but also through sexual intercourse, should we try to prevent them from harming others by discouraging them from sexual activity? It is difficult to see how this could be done through the law, except by the draconian step of isolating or incarcerating AIDS carriers. Society is unlikely to insist on this most dramatic interference with individuals' rights, even to stop a killer disease like AIDS. On the other hand, if AIDS devastates a large proportion of the population, who knows what steps would be taken, or should be taken, to counter it?

Although legal restrictions would seem inappropriate to many who value autonomy, an advisory advertising campaign would upset few and would achieve more. Here we are clearly within the realm of friendly persuasion rather than coercion. The provision of information on how to detect symptoms of AIDS and on why it would be dangerous for the partner if an AIDS carrier engaged in sexual activity, would generally be regarded as an obvious and acceptable response to the disease. Imprisoning all homosexuals in solitary confinement, at the other extreme, would generally be regarded as an over-reaction to the suggestion that AIDS thrives on homosexual practice.

Nevertheless, AIDS has probably had some effect on the way in which some heterosexuals regard homosexuals. Whether or not this distaste is an irrational prejudice, it is only one of many dislikes or preferences which we dub discrimination. Apart from sexual orientation, discrimination on grounds of gender, colour, nationality, race or religion has a long history.

But why is discrimination A Bad Thing? Discrimination used to be a good word. To say of a music critic that he is discriminating is a compliment. To discriminate between good and bad music is his job. Discrimination means choosing. But in the context of racial discrimination against Blacks or Jews, for example, discrimination is A Bad Thing because the choosing is based on irrelevant criteria. Race is held to be irrelevant to most activities and therefore racial discrimination is immoral.

Religious discrimination could fall under the same criticism although, interestingly, British legislation against race and sex discrimination is not matched by laws against religious discrimination.

But society rests on some forms of discrimination. We discriminate on grounds of talent. Few would condemn a merchant bank for discriminating between a brilliant financier and the innumerate penniless spendthrift when selecting a chief executive. Some people of course would say it is not fair to discriminate between the two individuals in terms of the financial benefits they receive from society but many more would accept even that discrimination. Certainly the discerning choice of the former for the role of chief executive is obvious.

The important question is: which grounds of discrimination are relevant and which are irrelevant? Returning to the old hobby-horse, we must ask, for instance, whether discrimination against homosexuals is morally acceptable and should be countenanced by the law. We have already mentioned a Scottish case which showed the courts allowing discrimination against homosexuals in employment More recently American courts have considered the case of James Dronenburg who served nine years in the US Navy as a linguist and cryptographer with top security clearance.[1] He was discharged in accordance with a Navy instruction that any member of the Navy 'who solicits, attempts or engages in homosexual acts shall normally be separated from the service. The presence of such a member in a military environment seriously impairs combat readiness, efficiency, security and morale'. Discrimination on grounds of sexual orientations, then, is still a contentious issue in legal systems on both sides of the Atlantic.

So, simply intoning the concept of equality or freedom from discrimination is not enough to provide a basis for law. We need to know which types of discrimination are beyond the pale. We also need to decide *exactly* why some types of discrimination are unacceptable.

Otherwise a problem can arise in considering reverse discrimination. This is also called positive discrimination or affirmative action. It means discrimination in favour of a previously oppressed group. If we think that discrimination *simpliciter* is wrong, then even reverse discrimination is wrong. But if we think why the normal form of discrimination against, for example, Blacks is wrong it is because it takes account of a criterion which is irrelevant. In the example of race discrimination: direct discrimination against Blacks, as if they were of inferior moral worth to Whites, denies them opportunities and benefits. Blacks may be denied places at college, for example, just because they are black.

A scheme of reverse discrimination on the other hand would make available some college places to Blacks in preference to Whites. But this is not because Whites are considered morally inferior. Race is not irrelevant here because what the society is trying to do is to improve the conditions of a race which has historically been oppressed. This is not to say of course that those Whites who miss out on college places will not feel annoyed.[2] There is no easy solution to the problems caused by past discrimination. If we merely stop discriminating now we may not have done enough to rectify our ancestors' injustices, but if we do more in the form of reverse discrimination some people will be paying for the sins of their forefathers or even other people's forefathers. Society has to make a delicate judgement.

It is my purpose here merely to observe that much of contemporary politics is based on conflicting notions of the relevant criteria for discrimination. To put it another way, even if we accept that like cases should be treated alike we have to decide what counts as like cases and how unalike our treatment should be of unalike cases. Government involves benefiting some sections of society at the expense of others. I don't mean that in any derogatory sense. Government might well take from the rich through progressive taxation to redistribute to the poor. It could on the other hand do the reverse, perhaps in more subtle ways, by taking from the poor and giving to the rich. But either way it is engaged in an exercise of discrimination or treating people unequally even if the ultimate aim is to achieve some vision of equality. So much of politics is about these moral values of equality or freedom from discrimination, whether that is dressed up as being about morality or being about justice or more cynically being about politics. We need to broaden our horizons from the preoccupation with criminal law bans and some forms of sexual behaviour to encompass the full range of ways in which we treat different groups in society.

If anybody thinks that this is a simple matter, two recent problems in British politics should give them cause for thought. Firstly, should our legal system recognise a right in Muslims to run their own schools with state help? We allow Catholics and Anglicans and Jews to do this. If we do not allow Muslims the same facility is this not religious and perhaps racial discrimination? On the other hand we think that Muslim values place women in a position, to non-Muslim eyes, of inferiority. So if we allow Muslim schools in order to rebut the charge of religious or racial discrimination we may open ourselves up to the charge of sex discrimination. Animal rights activists see another dilemma in the question of whether Muslims, or Jews for that matter, should be allowed their ritual slaughter of animals for food when others consider that to be immoral treatment of those animals.

A second contemporary problem concerning race discrimination has arisen in the controversy over social workers' placing of black children for fostering. Should black children only be placed with black families? If that is the policy, as it appears to be, of certain local councils, is it an acceptable form of racial discrimination or would problems of black children with white foster parents instead be a form of racial discrimination?

Without attempting to resolve these thorny problems I hope to have suggested that whatever the solution, society *has* to be concerned with moral values. When it chooses to use the law over other techniques in combating discrimination, or indeed in entrenching discrimination, then again the law has to rest on moral values. The question whether the law should enforce morality is a redundant one. The question ought to be instead *which* morality should the law enforce or, since I have tried to move the argument beyond legal enforcement, which morality should society reflect and encourage?

Liberty: Freedom of Expression, Pornography

13

Liberty is another value which seems to lie at the heart of our concern. We have considered how this concept is developed in different ways. An understanding of autonomy has been significant in many of our discussions. But it is fitting to reflect particularly on one subset of liberty: freedom of expression. This slogan has been especially dear to many theorists. As far as we are concerned, part of our approach is to emphasise the value of discussion about moral controversies. But like equality and all other values, freedom of expression has to compete with other rights in our society. We need to know why it is important. We need to decide its limits.

Pornography is one example of a law and morals controversy in which freedom of expression is often invoked. Pornography is difficult to define but, so I am told, easy to identify. Perhaps the reader may find it useful to relate the arguments which follow to specific forms of 'entertainment', such as

1. a live sex show, where actors copulate on stage
2. a 'hard core' video which shows explicit scenes of a violent gang raping a young victim
3. a 'soft core' magazine available on the shelves (albeit the upper shelves in an attempt to discourage children and short adults from reaching it) of a respectable newsagent
4. page three of the *Sun*, Britain's top-selling newspaper, which includes daily a photograph of a bare-breasted young woman.

Are any or all of these pornographic?

Whether or not we classify them as pornography, how should we decide on the law's reaction to these phenomena? The harm-to-others test would focus attention on whether or not each example 'harms' someone. In each case the someone might be a participant in the display, a viewer, somebody who suffers an attack induced by the example and perpetrated by a viewer, or somebody who suffers less dramatically through the attitudes adopted towards her by the viewers.

The last two suggested victims lead on to a consideration of the 'harm' involved. There might be physical harm cause to Z by X after X has seen Y's video. X might watch Y's rape scene and then imitate it by raping Z. If such a causal link could be established and if X would not have committed rape otherwise, we could talk of Y's video, as well as the individual X, causing physical harm to another. This might merit intervention under the harm-to-others approach although, as we have noted, in reality there would be other questions to ask. Some would ban a video if it could be shown to have caused one rape. Others would say that the 'benefits' of the video and the rights of non-rapist viewers and producers have to be weighted in a balance. Moreover, many would question whether a legal ban would be effective. Might it not simply drive the pornography underground, making it perhaps even more perverted and perversely more attractive to potential viewers?

The pornography might involve a moral 'harm', if by that is meant a diminution of one's powers of discerning good and evil and if making or viewing pornography has such an effect. But is moral degeneracy, even if that is proven here, a 'harm' for the purposes of the harm-to-others test? If it were, then the principle would permit legal intervention in a wide range of cases which have hitherto escaped the law's condemnation.

A third sense of harm which might be involved is the offence caused to non-participants who feel disgusted at the knowledge that pornography is available. Again, if this is allowed to count as 'harm', the law's ambit will be very wide indeed. It is precisely this kind of offence which Hart and Mill were seeking to *exclude* from the justifications for legal intervention. Devlin, on the other hand, would regard such offence as a legitimate ground for legal action if the offence was so great that it caused widespread 'intolerance, indignation and disgust'.

I use the phrase 'legal action' to warn that legal intervention does not have to involve outright bans. If people are offended through seeing pornographic photographs on the covers of magazines in newsagents, that may very well be a good reason for placing restrictions on the cover or making the pornography only available inside sex shops. In these ways the passer-by does not have to witness pornography but the pornography is not prohibited.

The Williams Committee on Film Censorship and Obscenity was sympathetic to that line of thought and recommended restrictions on the accessibility of pornography so as to protect objectors from offence through witness. But the Williams Committee was not prepared to accept offence through knowledge as a reason to ban pornography. This takes us back to Mill and Hart.

Another way in which pornography may harm others can be described as more pernicious or more nebulous than offence through knowledge, depending on one's point of view. The argument is that the victims of pornography are generally women. Pornography affects the way in which men regard women and indeed the way in which women regard themselves. Pornography degrades women, treating them as inferior to men and as sex objects. Not only may that influence the consumers of pornography to treat women in an unacceptable way, but also it may contribute to a climate in which even non-consumers of pornography have their attitudes influenced.

Whether or not page three of the *Sun* is pornographic, it provides a good test for this last point. The British public is well aware of what appears on page three of the *Sun*, even if much of that public does not buy the *Sun*. Page three has become famous or notorious, a joke or a pleasure or a disgrace. But does it affect our respect for women?

When I have discussed this with students, some men claim that page three has no impact at all on their attitude towards women. Other men and many, but not all, women students seem to find the notion of page-three girls offensive and degrading. They feel that page three contributes to an unhealthy attitude. It is not only some young men who find this hard to accept. So do older self-proclaimed 'liberals' who are stuck in a time-warp. When they were students in the late 1950s and early 1960s, it was all the rage to stand up for freedom of expression including pornography. They find it hard to adjust to an age in which other values, such as equality for women, are challenging their dogma.

So what if some women do object to page three? Well, again it depends on a large range of factors but if enough people object and if enough *Sun* readers or potential readers in particular object, I suppose that the *Sun* would stop page three voluntarily. There seems to be little sign that such a change of policy is imminent. Should the law force a change in advance of public opinion?

It would have become apparent that although we began considering this issue in terms of the harm-to-others principle, we are moving into my preferred framework. In order to decide what to do, we have to establish, as best we can, firstly, on the debit side of consequences, whether pornography does cause physical harm and whether it does contribute to a sexist climate, and on the credit side, whether pornography does, as is sometimes claimed, release or divert sexual frustrations so as to reduce the number of sexual attacks.

Then we must establish, on the basis of those facts or hunches, whether moral values like equality between, and respect for, the sexes and a 'proper' understanding of and respect for our sexuality require action to discourage or restrict pornography, or whether moral values like freedom of expression, privacy and toleration are more important considerations.

Now, we have so far concentrated on the objections to pornography. Some may reject as hypersensitive the argument I have ascribed to some female students. But can the defenders of pornography produce any positive arguments in favour of allowing freedom of expression?

There may be a problem in bringing pornography under the umbrella of freedom of expression. Or, rather, the attempt to include pornography may weaken the arguments commonly put forward to tolerate freedom of expression.

One explanation of the emphasis often placed on freedom of expression is that restrictions may stifle the emergence of the truth. Today's heresy may become tomorrow's orthodoxy but only if it is allowed expression. But it is difficult to imagine how pornography helps us to ascertain the truth.

A second justification is that freedom of expression is a vital adjunct to the autonomous life, to the development of one's own personality. This is the best argument in favour of the pornographers. But *is* unlimited freedom of expression really so necessary to our self-fulfilment?

A third rationale for regarding freedom of expression as important is that it is an essential part of the democratic process. Democracy involves some idea of citizen participation and a choice between political representatives. If we are to choose on the basis of what we might call 'informed consent', then we need to allow criticism of politicians and their policies. But, again, whatever the justification for allowing the pornographer freedom or for curtailing his activities, they cannot be the same as the argument for freedom of expression in the political arena. The pornographer's freedom does not contribute to the free flow of information and the free flow of criticism of government which we need to operate a fully fledged system of democracy.

Before coming to any conclusions about the value of freedom of expression, therefore, it is necessary to consider more than just pornography. We ought to look at freedom of expression in the political arena.

In the summer of 1985 the BBC had scheduled a *Real Lives* documentary about Ulster extremism. Under pressure from the Home Secretary, however, the BBC governors banned the programme. Or, at least, that was how it seemed initially. It soon became apparent that they had actually *postponed* it.

Now what does freedom of expression mean? Article 10 of the European Convention on Human Rights begins in uncompromising terms: 'Everyone has a right to freedom of expression', but then come the exceptions. Restrictions are allowed which are 'necessary in a democratic society, in the interests of national security, territorial integrity or public safety, for the prevention of disorder or crime, for the protection of health or morals, for the protection of the reputation or rights of others, for preventing the disclosure of information received in confidence, or for maintaining the authority and impartiality of the judiciary'. So there!

Freedom of expression is far from being absolute. We are not free to defame others, to utter obscenities in public, to incite racial hatred, to advertise on the BBC (at the time of writing), nor to advertise cigarettes on ITV (although we can advertise cigarettes on the BBC under the thin cover of sports sponsorship).

But should we be free to see *Real Lives*? In discussing the BBC ban, different individuals' freedom of expression comes into question. So whose freedom of expression most concerns us? There are many candidates. They may begin with the Prime Minister and the Home Secretary, include the BBC governors, management and programme makers, take in the Ulster extremists featured in the programme and end up with the rest of us since we may have wanted to watch the programme and talk about it.

Once we have established whose freedom of expression is at issue, the next query is whether we are really concerned with their abstract freedom to speak or their practical opportunities to get their message across.

Freedom of expression becomes a powerful political weapon when the speaker is given a public platform by, for instance, the BBC. But freedom of expression does not *require* the BBC to give particular groups the opportunity to express their views. The vital questions for a democracy, given freedom of speech, are: what should be the criteria governing the granting of such opportunities and who should set them? Should it be up to producers or managers or governors or government or governed?

If we are championing freedom of expression, I don't see why the BBC should object to the Prime Minister or the Home Secretary or anyone else venturing an opinion. The freedom to voice one's views is, after all, what we are meant to be defending. That is not to say, of course, that cabinet ministers are just like anyone else. The views of the minister responsible for broadcasting, the Home Secretary, are bound to be more influential than ours especially when the future of the BBC seems uncertain. But it is perhaps inevitable that the ultimate power of veto on broadcasting rests with the Home Secretary since so many of those restrictions on freedom of expression concern the national interest. The Home Secretary may well be best placed to make the ultimate decision because he is elected, accountable and removable. Hence the law gives the Home Secretary a veto power over the BBC.

That is not to say he should have used it here if the BBC governors had rejected his request to ban the *Real Lives* programme. The national interest may restrict freedom of expression in some cases but free speech is still the general principle and there is a strong presumption in its favour. The BBC and the Home Secretary can only interpret the exceptions adequately if they have some theory as to the purpose behind the normal rule of free speech. Why is it valuable? One standard argument is that it is only in a climate of liberty that we can test the truth of various notions. If we suppress today's heresy we may never discover tomorrow's truth. Another aspect is that we have a right to informed self-determination. A further point is that democracy depends on criticism. If we find these rationales convincing, we may still accept that freedom of expression is a limited right. It may conflict with other rights, say national security or one of the other exceptions in the European convention. Nevertheless, many regard it as a fundamentally important right at the political level.

But if our justification is really that freedom of expression is valuable instrumentally, in so far as it contributes to a healthy democracy, then it is presumably defeasible if it works against democracy in a particular instance.

This raises the real question of principle to emerge from the *Real Lives* debate: does the democracy of the United Kingdom have the right to suppress free speech where that free speech is advocating destruction of the UK's democracy?

Again, the best method of analysis is to separate the predicted consequences from the moral values which will determine our response by asking, would showing *Real Lives* at that particular time cause 'harm' in any way? Would it inform the public of the attitudes adopted by extremists in Ulster? Would that lead the public to sympathise with one or other extremist? Or would it expose the extremists and alienate public support? Against this, we would have to pose the question, *exactly* why is freedom of expression in this instance regarded as valuable?

Finally, what counts as 'censorship'?

Apologists for the BBC governors argue that the Home Secretary's intervention amounted to censorship even though he did not exercise his formal veto power. But in a world of free speech we must expect others to challenge our views and thereby attempt to change them with varying degrees of pressure, ranging from a parental withering look to a Home Secretary's withering letter. There is no need to respond to such pressure by lying on our backs, kicking our legs in the air and whining 'censorship'. It is possible to stand firm and challenge the Home Secretary to put his legal power where his mouth is. It is possible but it is also difficult. What is vital, however, is for all of us to examine our commitment to freedom of expression, to work out why we value it, and to decide how much pressure we can legitimately apply to persuade others to express 'freely' the views we ourselves hold.

The Morals of Law

14

The debate about the enforcement of morality through law has often actually been a debate about the criminal-law banning of sexual *im*morality. A wider perspective is needed. We ought to progress from sexual immorality to consider the whole range of morality or, if one regards this virtue as broader, the whole scope of justice. We ought also to widen our sights beyond the criminal law to other legal ways of registering disapproval and, in the context of seeking justice, of positively promoting a moral basis of, say, greater equality in society (or whatever our goal might be). To gain a well-rounded view of the strengths and weaknesses of law as one technique by which to achieve this, we need an even broader vision, encompassing all the other ways in which we influence each other's behaviour.

That task is enormous and well beyond the scope of this brief book. But I hope that this is a small start to the process of recognising the issues at stake. Let us move on from preoccupation with alleged differences in approaching the redundant question whether the law should enforce morality, by accepting that of course the law does and should depend on moral values. Then we can concentrate on arguing about *which* values the law should accommodate and how much emphasis should be given to, say, the liberal moral values of tolerance and autonomy. We need to test our moral intuitions or assumptions, our ideas of how they should apply to particular disputes and our predictions of the likely consequences of the legal or extra-legal options.

Set-piece controversies featured by the media, such as the Warnock or Gillick sagas, offer the perfect opportunity for testing our views. Day after day other issues rise and fall as media attractions but these too need consideration. Against a background of free expression (albeit ultimately a limited freedom), we can all learn and contribute to the debates. But we need expert help in assimilating the information on which to base our thoughts. The structures of law reform need careful attention. In the absence of such expert assistance, we rely desperately not only on a genuine freedom of expression via the media but also on genuine *access* to the media for those who can contribute to our understanding.

There will still be irreconcilable differences of opinion on all the key issues: the moral values themselves, their application and so on. So long as the facility exists for people to reach a reflective, informed conclusion, I think we should be loath to distrust the views of the people. Liberals have sometimes seemed scared to trust their fellow citizens and so have busied themselves in telling them not to be moralistic or paternalistic. In this way, some supporters of autonomy seek to restrict the autonomy of those who disagree on the priority of autonomy! In fact, they are being moralistic and paternalistic themselves. Now there is nothing inherently wrong in being moralistic or paternalistic. On the contrary, but for their pejorative overtones, we might be ready and willing to adopt these attitudes.

But the question is *when* and *how* to support *which* moral values. We should take advantage of the 'Great Debates' to argue about morality. We should also take advantage of the convention about some issues being 'votes of conscience' to inform our MPs about our visions of morality. And once the public has tasted the involvement it has felt over the Warnock Report, there is no reason to restrict ourselves to a narrow range of issues on which our views are accepted as counting. When votes are on party lines, of course, we have to direct our energies to the party platforms rather than MPs. But the principle remains the same. The law and morality debate is not for intellectuals who distrust the feelings of the rest. It is for everyone and it is up to the intellectuals to contribute to the examination of those feelings.

Let us be more specific about what I hope has emerged from this book. It has not been my intention to provide answers but to raise questions. Those questions should encourage the reader to exercise his or her autonomy and to think about the answers and about further questions.

What, then, are the morals to be drawn from the moral dilemma raised by the relationship between the law and morals?

The Scope of the Debate

Law and morals is a convenient phrase which comes at a middling level of generality. More narrowly, the debate has sometimes concentrated overmuch on criminal law and sexual immorality. More widely, law and morals should be located within the full spectrum of ways in which our whole range of values may be supported, enforced, reinforced or undermined. Incidentally, it will be apparent that, despite my protestations to the contrary, sex has constantly reappeared in our discussions. Mary Warnock explains this well when she says,

I do not believe that there is a neat way of marking off moral issues from all others; some people, at some times, may regard things as matters of moral right or wrong, which at another time or in another place are thought to be matters of taste, or indeed to be matters of no importance at all. But it seems likely that in any society, at any time, questions relating to birth and death and to the establishing of families are regarded as morally significant.

I hope that those dilemmas which clearly involve moral values will whet the appetite of us all to re-examine the values which we support across the whole range of moral, social and political problems.

I also hope that we will not shelter behind the issue of what the *law* should do. That is an important question and, perhaps because it can be discussed impersonally and without direct impact, it often provides one spur for reflection. But even if our views on the law's proper response to moral questions may not be influential, we may still have to respond to moral dilemmas personally. Whatever the law says about discrimination, the heterosexual white male ought to examine his *own* attitudes towards homosexuals, Blacks and women (and indeed to other heterosexual white males). Even where the law is unlikely to protect a group from discrimination we should examine our attitudes towards others. How do we treat the people we meet and are we prejudiced against those poorer than ourselves, or richer, older, younger, cleverer, more bizarrely dressed, more outrageously coiffured? And if we are, so what?

Moral number one, therefore, is to use the issue of law and morals as a catalyst for reflection on our whole range of personal and public responses to the whole series of problems which involve value judgements.

Stereotyping

It is easy to typecast opponents or imagined opponents and easy to make false assumption in such stereotyping. Thus Catholic views on law and morals should not be dismissed just because they are Catholic and 'we all know what Catholics think'. We probably do not know. All Catholics do not think alike. In any event, the Catholic tradition might have something useful to offer even non-Catholics by way of approach to law and morals. At the very least, it will be easier to respond to Catholic arguments if one understands them.

Similarly, it is currently, although doubtless only temporarily, fashionable in some quarters to decry liberalism. Liberalism is regarded as a wishy-washy hippy left-over from the 1960s. But, again, the liberal philosophy is more often criticised, or indeed invoked, than it is actually understood. All liberals do not think alike. Indeed, the disagreements between liberals are legendary. Non-liberals need to know why liberals affirm certain values. Non-liberals who try to understand what liberalism has to offer may find that they are not non-liberals after all.

Moral number two, therefore, is to discover what motivates those who disagree with us and not to ignore opposition or deride it on the basis of misconceived stereotyping.

Harm to Others

It is easy to incant a magic formula, more difficult to apply it, and very difficult to understand why it should be the solution to all our problems.

Moral number three, then, is that while Mill's harm-to-others principle may seem attractive, it does not do the thinking for us on law and morals: we need to decide what counts as harm, who counts as others and what else needs to be considered.

Factual Predictions/Moral Evaluations

Instead of relying on Mill, therefore, it is more helpful to admit that there is no one simple principle on which to rely.

Moral number four, then, is that hard choices have to be made and that there is no escape from analysing the contrasting factual and moral assumptions which characterise disagreements on law and morals.

Structures for Decision-making

It follows that a lively, open debate is most helpful. If different interested parties explain their evidence for their predictions of the consequences of legal options, we can have more confidence that wildly erroneous assumptions will not survive the examination.

Good quality research by pressure groups (for example, the National Council for Civil Liberties), independent bodies (for example, universities) and government-inspired committees or government departments, may help rid us of misconceptions.

It seems to me that a court is particularly unsuitable for this exercise, as in the Gillick case. *No* evidence was adduced at all. The differences between the Law Lords reflected partly their different hunches as to what would happen if the DHSS or Mrs Gillick succeeded.

Nor did the Warnock Report concentrate on providing or sifting information on *in-vitro* fertilisation. It did not provide any real support for the ideas that surrogate motherhood might have a detrimental effect on all or any involved. Nor did it examine the likelihood of improvements in medical science arising from experiments on embryos but which would not arise from alternative scientific endeavour.

The pornography and TV-violence debates are notorious for the difficulty of establishing what the consequences might be.

I do not believe that it *is* possible to predict or calculate all the consequences of our actions in advance. That is one of the flaws of utilitarianism. But that does not mean that we should abandon all interest in the enterprise. In many disputes over law and morals, as we have seen, the heat could be taken out of the argument if the protagonists considered whether their proposed legal measures would be counter-productive. This is one of Mrs Gillick's errors, according to her opponents.

Once the alternatives have been clarified so far as is possible, then we have to consider the moral values at stake.

Mary Warnock herself claims that 'there is no such thing as a moral expert'. But surely some people are better than others at spotting illogical, incoherent and inconsistent arguments. Surely some people are better than others at understanding theories of morality such as utilitarianism or a rights-based doctrine. And, most importantly, we can all become more expert ourselves by study of, reflection on, and practice of discussions about morality. Warnock is right to say that 'Everyone's conscience is his own'. But the point of moral expertise is not necessarily to convert everyone's consciences into holding the same set of values. It is to help us refine and develop our own consciences.

In my view, therefore, 'moral experts' could have a useful role on committees of inquiry. Apart from expertise, we do not all have the time to study the moral dilemmas in great depth. We can be helped by the experts clarifying the alternative moral arguments for us.

As regards both facts and moral values, then, I believe that there is a role for 'Super-Warnocks'.

But once the issues have been thus clarified, it is for all of us to express our views, if we so wish. Moreover, all our views should count. If 'liberals' distrust the moral values of others they should concentrate on educating the misguided into the 'right' values. It is a risky short-cut to argue that other people's views should not count.

It is, of course, also risky to *allow* all views to count. The majority's opinion is not sacrosanct. It should be subjected to the rigorous scrutiny of 'critical' morality. And, in practice, the process of legislation tends to filter out the most unpalatable views.

But, in principle, it must be preferable to *educate* the majority into not abusing its power. Democracy rests on a tension between respect for the wishes of the majority and respect for the interests of minorities who might fall foul of majority oppression. The long-term solution is for the majority to come to respect the interests of the minority. Discounting the views of the majority in the legal enforcement of morality may be a short-term ploy but changing the *attitudes* of the majority is a better response. Of course, a change in the law *may* lead to a change in attitude. Laws against racial discrimination have an educative role. But it is the underlying attitudes which ought to be the focus of attention.

The problems of democratic decision-making themselves deserve a book rather than a few paragraphs. For present purposes however, we should at least recognise the argument that 'the majority thinks X should be banned'. This is not conclusive to a discussion of law and morals. But neither is it completely irrelevant. In most issues '*the* majority' can be broken down into those who think X should be restricted, severely restricted, banned with a minor penalty or banned with serious deterrents to enforce the prohibition. Even if there were a majority, according to opinion polls, for one particular response, that majority might change as it becomes educated about the factual and moral disputes underlying the dilemma. If the reflective view of the majority is united in favour of one course of action, some visions of democracy would still allow the powers-that-be to resist majoritarian pressure. On this approach, the government is entitled to lead, rather than follow, public opinion. But *if* these conditions have been fulfilled, we would be sceptical of any government which could not win the argument, nevertheless enforcing its views. In practice, unfortunately, these conditions have rarely been achieved. That is where we have been going wrong. The time has come to focus on developing the moral education of us all. Through the mass media, we all follow controversies over law and morals.

The media and the public have invaluable contributions to make to the resolution of those controversies. But first the public needs the opportunity to reflect on the moral dilemmas in an informed and educated way.

Moral number five, then, is that we need to structure democratic decision-making to assist the clarification of factual and moral disputes; experts should help all of us make responsible decisions on law and morals through educating us instead of taking the decisions themselves for fear of our untutored responses.

Hints on the Relevant Moral Values

At the core of many disputes has been the value of autonomy, the conditions necessary for autonomy to flourish, and the circumstances in which paternalism should be invoked. There are various ways of giving effect to our concern for autonomy.

First, one could develop Mill's harm-to-others principle.

Second, one could adopt the limiting principles set out by Devlin as moral values which always ought to feature: toleration and privacy should be respected (but not always allowed to override all other considerations) and the law should aim to establish minimum not maximum standards of behaviour.

Third, one could adopt Hart's approach that only the 'universal values' merit legal support and not those which fluctuate according to fashion (he is thinking of sexual morality), unless of course 'harm' is being caused to others or physical harm is being self-inflicted.

For all the controversy, each of these schemes achieves much the same effect of differentiating law and morality. Each places hurdles in the way of those who would leap from the view that something is immoral to the conclusion that it should be illegal.

But each theorist will also have influenced our views on what is immoral in the first place. Taking morality in a wide sense, as we have done, our consideration of various issues should have led us to question our moral and political values.

We will either have renewed faith in those values or a refined or radically altered idea of their significance. By testing the consistency of our moral intuitions over a range of social problems, we should have a greater appreciation of our own values and the certainty or uncertainty with which we hold them.

In particular, we should examine our conceptions of liberty and equality and above all our understanding of the significance of autonomy. Rather than regarding 'paternalism' as beyond the pale, we need to consider how 'parental paternalism' can help us to develop as autonomous beings, when that ought to stop, and when medical or legal paternalism (as in the case of discouraging the initial taking of heroin) can help even adults to lead their lives as autonomously as possible.

Moreover, reflection on these matters may help us clarify how we wish to exercise our own autonomy. Our idea of the good life may be affected by thoughts about the value-conflicts we have examined. Although the law may generally concentrate on a minimum standard of behaviour, assessment of the morality of a dilemma should affect our own desire to go beyond the minimum as a matter of personal choice.

It is our own choice whether, within the law, we indulge in pornography, under-age sex, *in-vitro* fertilisation, acts of charity and so on. It is our own choice, within the legal constraints, how we think of the other sex, other races, other religions, other sexual orientations.

I slipped charity into the preceding paragraph because I want to end on a non-legal (but not illegal) note. Rock music appeared briefly at the beginning of this book, under threat from the Moral Majority. Now it can resurface as the epitome of moral worthiness.

Bob Geldof and the Band Aid–Live Aid ventures built on the work of the BBC in highlighting the dreadful Ethiopian famine in the mid-1980s. The pop world unleashed a tremendous force for goodwill and concern for our fellow human beings. Not only were millions of pounds raised to help alleviate suffering and begin to create the conditions for autonomous life, nor was it simply a fund-raising exercise — it was also a raising of consciousness. Even the words of the Band Aid Christmas record have a moral dimension. The rock stars enjoin us to 'pray for the other ones' and remind us that 'there's a world outside your window'.

The law and morals debate is a less spectacular introduction to the idea that there is a world outside our window. Its value may rest primarily in encouraging us to look out of that window. We do not always have to see there a *legal* solution to our problems. Indeed, the great virtues such as love and charity, exemplified by the Band Aid spirit, form part of our behaviour which is above and beyond the minimum standard we are told the law aims to enforce.

Our final moral, then, is to repeat our first moral: we should use the issue of law and morals as a catalyst for reflection on our whole range of personal and public responses to the whole range of moral dilemmas.

If law and morals are to prosper, we should all contribute to discussions on them, and action about them, as best we can. This introductory book now concludes with some suggestions for further reading. The way forward is to combine that reading with further argument both in private (at home, school, work or social gathering) and in public (through writing letters to the press and to MPs and by more ambitious campaigning). I hope that the reader has been encouraged to continue the debate.

Notes on Sources

1 The Law and Morals Debate

1 See the slim volumes by H. L. A. Hart, *Law, Liberty and Morality* (Oxford, 1968) and by Lord Devlin, *The Enforcement of Morals* (Oxford, 1968). Basil Mitchell adds to the debate in *Law, Morality and Religion in a Secular Society* (Oxford, 1968). I should make it clear that both Hart and Devlin were well aware of the importance of issues other than sex and the criminal law. A glance at the chapter headings of Devlin's book will indicate that. Nevertheless, the debate seemed to get stuck, naturally enough, on the immediate issues raised by the Wolfenden Report's recommendation that homosexual practice in private between consenting adults should no longer be a crime. We return to the Hart–Devlin debate in ch. 6.

2 *Saunders* v. *Scottish National Camps Association* [1980] Industrial Relations Law Reports 174 and [1981] IRLR 277.

2 Moral Majority? A Catholic View

1 Thomas Aquinas, *Summa Theologiae* (Blackfriars ed., Eyre & Spottiswoode, 1963–75, vol. 28, Thomas Gilbey OP) 1a2ae, question 96, article 2, reply.

3 Immoral Minority? A Liberal View

1 H. L. A. Hart, *The Concept of Law* (Oxford, 1976) ch. 9.

2 For a detailed discussion of autonomy and liberalism see Joseph Raz, *The Morality of Freedom* (Oxford, 1986).

4 Amoral Law? A Lawyer's View

1 *Donaghue* v. *Stevenson* [1932] Appeal Cases 562.

2 P. S. Atiyah, *An Introduction to the Law of Contract* (3rd ed., Oxford, 1981) pp. 2–3 and see the rest of ch. 1 for an excellent introduction to the values underlying the law of contract.

5 Mill's 'Harm-to-Others' Principle

1 John Stuart Mill, *On Liberty* (1859) now available in many editions, e.g. Oxford Paperbacks (1975).

2 Samuel Brittan, *Two Cheers for Self-Interest* (Institute of Economic Affairs, 1985), p. 15.

6 The Hart–Devlin Debate

1 See ch. 1 note 1 (above) and the suggestions for further reading on this chapter (below). I try to summarise only what I regard as the most helpful aspects of this debate and thus ignore the controversy over what is meant by 'society'.

7 A Different Framework

1 Jeremy Waldron, 'Rights and Trade-offs', *Times Literary Supplement*, 8 November 1985, p. 1269.
2 Archbishop of York, Letter to *The Times*, 3 June 1985.

8 Embryo Experiments and Surrogate Motherhood: Warnock

1 12.2. A test in which human sperm may fertilise hamster eggs is already used in the investigation of male subfertility. Men whose sperm will fertilise a specially treated hamster egg may eventually father a child, whereas those whose sperm will not are probably infertile. Although in the hamster test any resulting embryo does not develop beyond the two cell stage, it is possible that other similar forms of trans-species fertilisation tests could be developed. Unlike the hamster test, such tests might result in an embryo which might develop for a considerable period of time. Both the hamster tests and the possibility of other trans-species fertilisations, carried out either diagnostically or as part of a research project, have caused public concern about the prospect of developing hybrid half-human creatures.

12.3 We take the view that trans-species fertilisation when undertaken as part of a recognised programme for alleviating infertility, or in the assessment or diagnosis of subfertility, is an acceptable procedure, subject to certain safeguards. Since the object is to assess fertilising capacity, we see no reason why any resultant embryo should be allowed to survive beyond the two-cell stage. We recommend that where trans-species fertilisation is used as part of a recognised programme for alleviating infertility or in the assessment or diagnosis of subfertility it should be subject to licence and that a condition of granting such a licence should be that the development of any resultant hybrid should be terminated at the two cell stage. Any unlicensed use of trans-species fertilisation involving human gametes should be a criminal offence.

2 *Warnock Report*, Expression of Dissent B.2. and B.3.
3 The Unborn Children (Protection) Bill, sponsored by Enoch Powell, MP. See *The Times* Parliamentary Report, 16 February 1985 and 9 June 1985.
4 The Warnock members are listed at the beginning of the Report.

9 Contraception, Children's Rights, Parents' Rights: Gillick

1 The full judgments can be found as follows: Woolf J. [1984] 1 All ER 365; Court of Appeal [1985] 1 All ER 533; House of Lords [1985] 3 All ER 402.

The other cases referred to are: *R* v. *D* [1984] 2 All ER 449; *Hewer* v. *Bryant* [1969] 3 All ER 449

10 Medical Paternalism: Informed Consent

1 The full judgments can be found at [1985] 1 All ER 643.

11 Resource Allocation: Kidney Dialysis

1 See *Lancet*, 19 January 1985, p. 176.

12 Equality: Discrimination

1 See Ronald Dworkin's discussion in the *New York Review of Books*, 8 November 1984.
2 See Ronald Dworkin's discussions of the American *Bakke* case, originally published in the *New York Review of Books*, 10 November 1977 and 17 August 1983, now chs 14 and 15 of Dworkin's *A Matter of Principle* (Harvard, 1985).

13 Liberty: Freedom of Expression, Pornography

1 After the text was written, Clare Short, the Labour MP, introduced a Private Member's Bill on this very topic of banning such newspaper pictures. See *The Times*, 13 March 1986, Parliamentary Report.

Further Reading

I hope that readers will be encouraged to read more detailed studies. Here are my suggestions for the next steps; the bibliographies in these books will in turn lead on to more complex analyses.

2. For Catholic views, including the Archbishops' Statement, see Austin Flannery O.P. (ed.) *Abortion & Law* (Dominican Publications, Dublin, 1983). For an alternative approach to abortion, see Jonathan Glover, 'Matters of Life and Death', *The New York Review of Books*, 30 May 1985. Two renowned philosophical articles are conveniently reprinted in R. Dworkin (ed.) *The Philosophy of Law* (Oxford, 1977): J. J. Thomson, 'A Defence of Abortion' and J. Finnis, 'The Rights and Wrongs of Abortion'.

3. For an introduction to liberalism, see Ronald Dworkin's contribution to Bryan Magee, *Men of Ideas* (Oxford paperback, 1982). Then read Ronald Dworkin, *A Matter of Principle* (Harvard, 1985), chs 8, 9, 10, 7, 11, especially ch. 8, 'Liberalism', and Joseph Raz, *The Morality of Freedom* (Oxford, 1986).

4. Anybody who is encouraged to discover more about law, and especially potential law students, should read two excellent introductions: P. S. Atiyah, *Law and Modern Society* (Oxford, 1983) and P. Harris, *An Introduction to Law* (Weidenfeld and Nicolson, 2nd ed., 1984).

5. There are innumerable paperback editions of John Stuart Mill's *On Liberty*. After absorbing the works which follow for ch. 6, the Mill devotee will be stimulated by C. L. Ten, *Mill On Liberty* (Oxford, 1980). Joel Feinberg's monumental four-volume study, *The Moral Limits of the Criminal Law* (Oxford, 1984–5) has begun with a comprehensive analysis of what should count as 'harm': *Harm To Others* and *Offense to Others*.

6. Wolfenden, *Report of the Committee on Homosexual Offences and Prostitution*, Cmnd. 247 (1957).
   H. L. A. Hart, *Law, Liberty and Morality* (Oxford, 1968).
   P. Devlin, *The Enforcement of Morals* (Oxford, 1968).
   B. Mitchell, *Law, Morality and Religion in a Secular Society* (Oxford, 1968).
   The original lecture by Devlin and the initial reply by Hart can also be found reprinted in R. Dworkin (ed.) *The Philosophy of Law* (Oxford, 1977).
   R. Dworkin, *Taking Rights Seriously* (Duckworth, 1978) ch. 10.

7. Those who wish to explore alternative approaches to morality will enjoy Jonathan Glover, *Causing Death and Saving Lives* (Penguin, 1977) especially ch. 1. Anthony Flew, *Thinking about Thinking* (Fontana, 1975) is an entertaining introduction. More difficult introductions include D. D. Raphael, *Moral Philosophy* (Oxford, 1981) and J. L. Mackie, *Ethics: Inventing Right and Wrong* (Penguin, 1977). Bryan Magee's dialogues with fifteen leading philosophers in *Men of Ideas* (Oxford Paperbacks, 1982) form a stimulating introduction to philosophy. Ronald Dworkin's contribution for example, is an excellent basis for appreciating the work of John Rawls, *A Theory of Justice* (Oxford, 1971).

The example of violence on TV is analysed in an excellent collection of articles in the *Listener*, 23 January 1986. Of particular interest to my point about factual disputes in law and morals is Barrie Gunter's contribution which casts doubt on the significance of the 700 pieces of published research which purport to link TV violence and violence in the real world.

8. The *Report of the Committee of Inquiry into Human Fertilisation and Embryology* under the chairmanship of Mary Warnock was originally published by HMSO in 1984. It has since been reprinted with additional material by Mary Warnock as Mary Warnock, *A Question of Life* (Blackwell, 1985). This title is interesting since the Report has been much criticised for failing to consider directly the question when life begins. The lay-out of the new version is misleading. Mary Warnock's additional material consists of a personal Introduction which is sandwiched between her original Letter to the Government and the Committee's Foreword, and her personal Conclusion which appears between the Dissents and the Appendix. This seems confusing and inappropriate. The new chapters are not the work of the Committee. Some of Mary Warnock's assertions as to what motivated Committee members might be disputed by her colleagues. What weight should we attach to her claim that, for example, 'the Inquiry agreed unanimously that they disapproved of surrogate motherhood', given the Dissent on surrogacy which stated that 'There are, we hold, rare occasions when surrogacy could be beneficial to couples as a last resort'?

Just before the Warnock Report was published, Ian Kennedy anticipated and countered the majority view on embryo experiments: 'Let the law take on the test tube', *The Times*, 26 May 1984.

Two expressions of Catholic views on Warnock are quoted in the text. These are the evidence submitted to Warnock and the response to the Report by the Catholic Bishops Joint Committee on Bio-Ethical Issues, entitled respectively *In Vitro Fertilisation: Morality and Public Policy* and *Response to the Warnock Report* (Catholic Media Office).

9. See Ian Kennedy's discussion of the issues and the High Court and Court of Appeal judgments in 'The doctor, the pill, and the fifteen-year-old girl: a case study in medical ethics and law', published in M. Lockwood (ed.) *Moral Dilemmas in Modern Medicine* (Oxford, 1985). That was written before the House of Lords hearing in which Ian Kennedy appeared as junior counsel for the DHSS.

10. On medical law and ethics in general, see Ian Kennedy, *The Unmasking of Medicine* (Paladin rev. ed., 1983).

11. See G. Calabresi and P. Bobbitt, *Tragic Choices* (Norton, New York, 1978) for a discussion of resource allocation.

12. See A. Honoré, *Sex Law* (Duckworth, 1978) ch. 4, D. D. Raphael, *Justice and Liberty* (Athlone Press, 1980), especially chs 1 & 9, and David Pannick, *Sex Discrimination* (Oxford, 1986).

13. On freedom of expression in general, see Eric Barendt, *Freedom of Expression* (Oxford, 1985).

On pornography, see the Williams Report, *Report of the Committee on Obscenity and Film Censorship* (Cmnd. 7772, 1979) and R. Dworkin's critique in *A Matter of Principle* (Harvard, 1985), ch. 17.

On TV 'censorship', see the articles on the 'Real Lives' dispute in the *Listener*, 8 & 15 August, 1985.

14. As will be apparent from the above, one can confidently expect the quality press to publish material on future disputes over law and morals.

Index